A Simple Approach to
Differential Calculus

I0620044

A Review and Self-Teaching Workbook to Extensively learn Limits, Functions, and Derivatives.

Adegboye Samuel

*Copyright © *2023 by Adegboye Samuel.*

All rights reserved. Printed in the United States of America. No part of this book may be used or reproduced in any manner whatsoever without written permission except in the case of brief quotations em- bodied in critical articles or reviews. The following rights goes to any teacher or parent who purchases one copy of this workbook;

For any information, correction or assistance you can send a mail to smart learning on kunlektrapub@gmail.com

TABLE OF CONTENTS

Chapter 1

CALCULUS

Introduction

This book will introduce you to a new branch of mathematics called Calculus. Calculus was established by Newton and Leibnitz in the 17th century. It concentrates on some aspects of mathematics which include;

- ✓ Limits.
- ✓ Functions.
- ✓ Derivatives.
- ✓ Integrals.
- ✓ Infinite series.

It is a branch of mathematics that involves the continuous change of a function.

Types of Calculus

The two types of calculus are;

- ✓ Derivatives (Differentiation – Differential Calculus)
- ✓ Integral calculus

This book volume will focus on differential calculus which helps to solve the rate of change of a function of a particular quantity.

Differential calculus solves for the rate of change of a function with respect to another variable. For a perfect solution, we use the derivative to find the minimum and maximum values of the function. Considering a cartesian graph that has its co-ordinates to be $x - axis$ and $y - axis$, calculus makes use of variable x and y, with corresponding changes in the variables $(x$ and $y)$ and the function $f_{(x)}$.

Functions

What is a function?

A function is a relationship between two sets that joins each element of the first set (domain) exactly one element of the second set(co-domain). It is mostly denoted by f, and $sometimes$ g and h.

A function is a mapping whose co-domain is the set of numbers. For example;

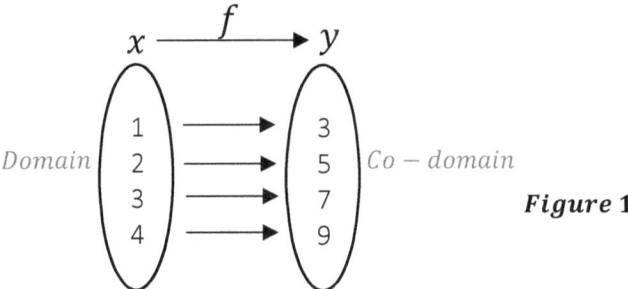

Figure 1

Both the domain and the co-domain are sets of numbers. The mapping f is a function. From *figure 1*, **3** is the f image of **1**. *Mathematically,*

$$f(1) = 3, f(2) = 5, f(3) = 7, and f(4) = 9.$$

The relationship which associates elements from x, to a unique image in the set y, is the addition of 1 to twice an element x; it produces the corresponding images in the set y.

Hence, if $y\epsilon Y$ is the $f - image$ of $x\epsilon X$, then we can write this as;

$$f: x \rightarrow 2x + 1 \text{ or } f_{(x)} = 2x + 1, \text{ or } y = 2x + 1.$$

Function Notation

When you have an equation in the form of;

$$y = 4x^2 + 3x - 7$$

We can deduce that it is a function of x. The form of the equation above can be written as

$f(x) = 4x^2 + 3x - 7$, and this is called **functional Equation.** when $x = 0$, $f(x) = f(0)$, and when $x = 1$, $f(x) = f(1)$.

Example 1: *If $f(x) = 3x^2 + 5x + 1$, find*

$f(1), f(2), f(3), f(-1), f(-2)$

Solution

$$f(x) = 3x^2 + 5x + 1$$

a. To find the value of $f(1)$, $x = 1$

$$f(1) = 3(1^2) + 5(1) + 1$$
$$= (3 \times 1) + (5 \times 1) + 1$$
$$= 3 + 5 + 1 = 9$$
$$f(1) = 9$$

b. To find the value of $f(2)$, $x = 2$

$$f(2) = 3(2^2) + 5(2) + 1$$
$$= (3 \times 4) + (5 \times 2) + 1$$
$$= 12 + 10 + 1 = 23$$
$$f(2) = 23$$

c. To find the value of $f(3)$, $x = 3$

$$f(3) = 3(3^2) + 5(3) + 1$$
$$= (3 \times 9) + (5 \times 3) + 1$$

$$= 27 + 15 + 1 = 43$$

$$f(3) = 43$$

d. To find the value of $f(-1)$, $x = -1$

$$f(1) = 3(-1^2) + 5(-1) + 1$$

$$= (3 \times 1) - (5 \times 1) + 1$$

$$= 3 - 5 + 1 = 9$$

$$f(1) = -1$$

e. To find the value of $f(-2)$, $x = -2$

$$f(-2) = 3(-2^2) + 5(-2) + 1$$

$$= (3 \times 4) - (5 \times 2) + 1$$

$$= 12 - 5 + 1 = 9$$

$$f(-2) = 8$$

Example 2: *Given $f(x) = 2x^2 - x + 5$, find*

1. $f(2) \div f(1)$
2. $f(2 + a) - f(2)$
3. $f(3 + a)$
4. $\dfrac{f(1+a) - f(3)}{a}$

Solution

$$f(x) = 2x^2 - x + 5$$

1. $f(2) \div f(1)$

   When $x = 2$

$$f(2) = 2(2^2) - (2) + 5$$

$$= (2 \times 4) - (2) + 5$$
$$= 8 - 2 + 5 = 15$$
$$f(2) = 11$$

When $x = 1$

$$f(1) = 2(1^2) - (1) + 5$$
$$= (2 \times 1) - (1) + 5$$
$$= 2 - 1 + 5 = 6$$
$$f(1) = 6$$
$$f(2) \div f(1) = \frac{11}{6}$$

2. $f(2 + a) - f(2)$

When $x = 2 + a$

$$f(2 + a) = 2(2 + a)^2 - (2 + a) + 5$$
$$= 2(4 + 4a + a^2) - (2 + a) + 5$$
$$= 8 + 8a + 2a^2 - 2 - a + 5$$
$$= 8 - 2 + 5 + 8a - a + 2a^2$$
$$= 11 + 7a + 2a^2$$
$$f(2 + a) = 11 + 7a + 2a^2$$

When $x = 2$, $f(2) = 11$

$$f(2 + a) - f(2) = 11 + 7a + 2a^2 - 11 = 7a + 2a^2$$
$$f(2 + a) - f(2) = 7a + 2a^2 = a(7 + 2a)$$

3. $f(3 + a)$

When $x = 3 + a$

$$f(3 + a) = 2(3 + a)^2 - (3 + a) + 5$$
$$= 2(9 + 6a + a^2) - (3 + a) + 5$$
$$= 18 + 12a + 2a^2 - 3 - a + 5$$
$$= 18 - 3 + 5 + 12a - a + 2a^2$$
$$= 20 + 11a + 2a^2$$
$$f(3 + a) = 11 + 7a + 2a^2$$

4. $\dfrac{f(1+a) - f(3)}{a}$

When $x = 1 + a$

$$f(1 + a) = 2(1 + a)^2 - (1 + a) + 5$$
$$= 2(1 + 2a + a^2) - (1 + a) + 5$$
$$= 2 + 4a + 2a^2 - 1 - a + 5$$
$$= 2 - 1 + 5 + 4a - a + 2a^2$$
$$= 6 + 3a + 2a^2$$
$$f(1 + a) = 6 + 3a + 2a^2$$

When $x = 3$

$$f(2) = 2(3^2) - (3) + 5$$
$$= (2 \times 9) - (3) + 5$$
$$= 18 - 3 + 5 = 20$$
$$f(3) = 20$$

$$\frac{f(1+a) - f(3)}{a} = \frac{6 + 3a + 2a^2 - 20}{a} = \frac{2a^2 + 3a - 20}{a}$$

Exercise 1

Given that $f(x) = x^2 + 4x - 7$. find

1. $f(0)$

2. $f(1)$

3. $f(2)$

4. $f(-3)$

5. $f(-4)$

6. $\dfrac{f(a-1)+f(2)}{2}$

Chapter 2

LIMITING VALUE OF A FUNCTION

The limit of a function is a fundamental concept in calculus, which is used to analyze the behavior of a function as it gets closer to a particular point. The primary focus is to resolve different problems on the Limit of a Function, which will make you get a better understanding of how to solve questions on it quickly.

Let us consider the function

$$F_{(x)} = \frac{2 - x}{3 + x}$$

The value of the functions above $F_{(x)}$ will get closer and closer to $\frac{2}{3}$ as $x \to$ 0 . that is, as x gets closer and closer to zero (0).

Then, it can be summarized

That $f_{(x)} \to \frac{2}{3}$ as $x \to$

Mathematically

$$\lim_{x \to 0} \frac{2-X}{3+X} = \frac{2}{3}$$

Essential Facts About Limiting Value Of A Function

1. If we consider the function

$$f_x = x^2$$

$$when \ x \to 2, f_x = 2^2 = 4$$

$$when \ x \to 2.1, f_x = 2.1^2 = 4.41$$

$$when \ x \to 2.01, f_x = 2.01^2 = 4.0401$$

$$when \ x \to 2.001, f_x = 2.001^2 = 4.004001$$

$$when \ x \to 2.0001, f_x = 2.0001^2 = 4.00040001$$

Observe the trend closely

Observed that $f_x \to 4$ _when_ $x \to 2$ from a value of $x > 2$ (2.0001, 2.001, 2.01, 2.1).

And also, from the function _when_ $x \to 2$ from the value of $x < 2$ (1.9, 1.99, 1.999, 1.9999), $f_x \to 4$.

$$f_x = x^2$$
$$when\ x \to 2, f_x = 2^2 = 4$$
$$when\ x \to 1.9, f_x = 1.9^2 = 3.61$$
$$when\ x \to 1.99, f_x = 1.99^2 = 3.9601$$
$$when\ x \to 1.999, f_x = 1.999^2 = 3.996001$$
$$when\ x \to 1.9999, f_x = 1.9999^2 = 3.99960001$$

Summarily.

$\lim\limits_{x \to 2^+} f_x = x$ is the notation for limiting value f_x as x approaches 2 from the right when $x > 2$. This notation is called a **_Right-hand limit of_** f_x as x tends to 2. $\lim\limits_{x \to 2^-} f_x = x$ is the notation for limiting value f_x as x approaches 2 from the left when $x > 2$. This notation is called a **_Left-hand limit of_** f_x as x tends to 2.

2. The limit can be directly found or can be found by using formulas.

Properties Of Limits

I refer to these properties of limit as the **laws of limit** because they enable us to evaluate the limits of a rather large class of functions without resorting to geometric figures or graphs.

Once you are familiar with these properties, the solution can be given to some complex problems on the limiting value of a function.

1. $$\lim_{x \to a} k = k$$

Where **a** and **k** are constant, the limit of a constant is the constant itself.

2. $$\lim_{x \to a} k f_x = k \lim_{x \to a} f_x$$

This implies that the limit of the product of a function and a constant is equal to the product of the constant and the limit of the function.

3. $$\lim_{x \to a}(f_x \pm g_x) = \lim_{x \to a} f_x \pm \lim_{x \to a} g_x$$

This implies the limit of the sum or difference of two functions is equal to the sum or differences of their respective limits.

4. $$\lim_{x \to a}(f_x \cdot g_x) = \left(\lim_{x \to a} f_x\right) \cdot \left(\lim_{x \to a} g_x\right)$$

(.) In mathematics implies **multiplication** or **product**. This implies that the limit of the product of two functions (it can be more than two functions) is equal to the product of their respective limit.

5. $$\lim_{x \to a}\left(\frac{f_x}{g_x}\right) = \frac{\left(\lim_{x \to a} f_x\right)}{\left(\lim_{x \to a} g_x\right)} \quad provided \quad \lim_{x \to a} g_x \neq 0$$

This implies that the limits of two functions are equal to the quotient of their limits, provided the limit of the divisor is not equal to zero.

6. $$\lim_{x \to a}(f_x)^n = \left(\lim_{x \to a} f_x\right)^n$$

where "**n**" is an integer.

7. $$\lim_{x \to a}\left(\sqrt[n]{f_x}\right) = \sqrt[n]{\lim_{x \to a} f_x} \quad where \ "\mathbf{n}" \ is \ positive \ integer$$

The laws of limit stated above will help when it comes to solving questions on limits

8. $\lim_{x \to a} \dfrac{1}{x} = 0$

Example 1: Evaluate $\dfrac{\left(\lim_{x \to 0} x^2 + 5x + 9\right)}{\left(\lim_{x \to 0} 2x^2 - 3x + 15\right)}$

Solution

$$= \frac{\left(\lim_{x \to 0} x^2 + \lim_{x \to 0} 5x + 9\right)}{\left(\lim_{x \to 0} 2x^2 - 3x + 15\right)}$$

Separate the functions using law 3

$$= \frac{\lim_{x \to 0} x^2 + \lim_{x \to 0} 5x + \lim_{x \to 0} 9}{\lim_{x \to 0} 2x^2 - \lim_{x \to 0} 3x + \lim_{x \to 0} 15}$$

Following law 2, separate the product of a constant and limit of a function

$$= \frac{\lim_{x \to 0} x^2 + 5\lim_{x \to 0} x + \lim_{x \to 0} 9}{2\lim_{x \to 0} x^2 - 3\lim_{x \to 0} x + \lim_{x \to 0} 15}$$

Replace x with 0, as $x \to 0$

$$= \frac{\lim_{x \to 0} 0^2 + 5\lim_{x \to 0} 0 + \lim_{x \to 0} 9}{2\lim_{x \to 0} 0^2 - 3\lim_{x \to 0} 0 + \lim_{x \to 0} 15}$$

To solve each limit, note that the limit of a constant is the constant itself,

$$= \frac{0 + 0 + 9}{0 - 0 + 15} = \frac{9}{15} = \frac{3(3)}{3(5)} = \frac{3}{5}$$

Therefore;

$$\frac{\left(\lim_{x \to 0} x^2 + 5x + 9\right)}{\left(\lim_{x \to 0} 2x^2 - 3x + 15\right)} = \frac{3}{5}$$

Example 2

Evaluate $\qquad \lim_{x \to 0}(7 - 2x + 5x^2 - 4x^3)$

Solution

$$\lim_{x \to 0}(7 - 2x + 5x^2 - 4x^3)$$

Separate the functions using law 3

$$= \lim_{x \to 0} 7 - \lim_{x \to 0} 2x + \lim_{x \to 0} 5x^2 - \lim_{x \to 0} 4x^3$$

Following law 2, separate the product of a constant and limit of a function

$$= \lim_{x \to 0} 7 - 2\lim_{x \to 0} x + 5\lim_{x \to 0} x^2 - 4\lim_{x \to 0} x^3$$

Replace x with 0, as x → 0

$$= \lim_{x \to 0} 7 - 2\lim_{x \to 0} 0 + 5\lim_{x \to 0} 0 - 4\lim_{x \to 0} 0$$

To solve each limit, note that the limit of a constant is the constant itself,

$$= \lim_{x \to 0} 7 - 2\lim_{x \to 0} 0 + 5\lim_{x \to 0} 0 - 4\lim_{x \to 0} 0$$

$$= 7 - 0 + 0 - 0 = 7$$

$$\lim_{x \to 0}(7 - 2x + 5x^2 - 4x^3) = 7$$

Example 3

Evaluate $\qquad \lim_{x \to 6}\left(\frac{x^2 - 36}{x - 6}\right)$

Solution

$$\lim_{x \to 6} \left(\frac{x^2 - 36}{x - 6} \right)$$

First of all, simplify the equation in the bracket

$$= \lim_{x \to 6} \left(\frac{x^2 - 36}{x - 6} \right)$$

Recall, $x^2 - a^2 = (x - a)(x + a)$

$$= \lim_{x \to 6} \left(\frac{(x^2 - 6^2)}{x - 6} \right)$$

$$= \lim_{x \to 6} \left(\frac{(x - 6)(x + 6)}{(x - 6)} \right)$$

Note,

$$\frac{(x-6)}{(x-6)} = 1$$

$$= \lim_{x \to 6} (x + 6)$$

$$= \lim_{x \to 6} x + \lim_{x \to 6} 6$$

Replace x with 6, as x $\to$ 6

$$= \lim_{x \to 6} 6 + \lim_{x \to 6} 6$$

To solve each limit, note that the limit of a constant is the constant itself,

$$= 6 + 6 = 12$$

Therefore;

$$\lim_{x \to 6} \left(\frac{x^2 - 36}{x - 6} \right) = 12$$

Example 4

Evaluate

$$\lim_{x \to a} \left(\frac{4x^3 + 2x^2 + x + 2}{x^3 + 2x + 10} \right)$$

Solution

$$\lim_{x\to\alpha}\left(\frac{4x^3+2x^2+x+2}{x^3+2x+10}\right)$$

Recall; $\qquad \lim_{x\to\alpha}\frac{1}{x}=0$

$$=\lim_{x\to\alpha}\frac{4x^3+2x^2+x+2}{x^3+2x+10}$$

Note; $\quad x^3+2x+10=x^3\left(1+\frac{2}{x^2}+\frac{10}{x^3}\right)$

$$4x^3+2x^2+x+2=x^3\left(4+\frac{2}{x}+\frac{1}{x^2}+\frac{2}{x^3}\right)$$

$$=\lim_{x\to\alpha}\frac{x^3\left(4+\frac{2}{x}+\frac{1}{x^2}+\frac{2}{x^3}\right)}{x^3\left(1+\frac{2}{x^2}+\frac{10}{x^3}\right)}$$

$$=\lim_{x\to\alpha}\frac{\left(4+\frac{2}{x}+\frac{1}{x^2}+\frac{2}{x^3}\right)}{\left(1+\frac{2}{x^2}+\frac{10}{x^3}\right)}$$

$$=\frac{\left(\lim_{x\to\alpha}4+\lim_{x\to\alpha}\frac{2}{x}+\lim_{x\to\alpha}\frac{1}{x^2}+\lim_{x\to\alpha}\frac{2}{x^3}\right)}{\left(\lim_{x\to\alpha}1+\lim_{x\to\alpha}\frac{2}{x^2}+\lim_{x\to\alpha}\frac{10}{x^3}\right)}$$

$$=\frac{\left(\lim_{x\to\alpha}4+2\lim_{x\to\alpha}\frac{1}{x}+\lim_{x\to\alpha}\frac{1}{x^2}+2\lim_{x\to\alpha}\frac{1}{x^3}\right)}{\left(\lim_{x\to\alpha}1+2\lim_{x\to\alpha}\frac{1}{x^2}+10\lim_{x\to\alpha}\frac{1}{x^3}\right)}$$

Replace x with α *, as* x $\to \alpha$

$$\frac{\left(\lim_{x\to\alpha}4+2\lim_{x\to\alpha}\frac{1}{\alpha}+\lim_{x\to\alpha}\frac{1}{\alpha^2}+2\lim_{x\to\alpha}\frac{1}{\alpha^3}\right)}{\left(\lim_{x\to\alpha}1+2\lim_{x\to\alpha}\frac{1}{\alpha^2}+10\lim_{x\to\alpha}\frac{1}{\alpha^3}\right)}$$

If $\lim_{x\to\alpha}\frac{1}{x}=0$, substitute into the equation above.

$$\frac{(4+2(0)+0+2(0))}{(1+2(0)+10(0))}=\frac{4}{1}=4$$

Therefore,

$$\lim_{x \to a} \left(\frac{4x^3 + 2x^2 + x + 2}{x^3 + 2x + 10} \right) = 4$$

Example 5

Evaluate $\lim_{x \to 1} (4x^3 + 3x^2 + 2x - 2)$

Solution

$$\lim_{x \to 1} (4x^3 + 3x^2 + 2x - 2)$$

Separate the functions using law 3

$$= \lim_{x \to 1} 4x^3 + \lim_{x \to 1} 3x^2 + \lim_{x \to 1} 2x - \lim_{x \to 1} 2$$

From law 2, separate the product of a constant and limit of a function

$$= 4\lim_{x \to 1} x^3 + 3\lim_{x \to 1} x^2 + 2\lim_{x \to 1} x - \lim_{x \to 1} 2$$

Replace x with 1, as x → 1

$$= 4\lim_{x \to 1} 1^3 + 3\lim_{x \to 1} 1^2 + 2\lim_{x \to 1} 1 - \lim_{x \to 1} 2$$

To solve each limit, note that the limit of a constant is the constant itself,

$$= 4(1) + 3(1) + 2(1) - 2$$

$$= 4 + 3 + 2 - 2 = 7$$

Therefore,

$$\lim_{x \to 1} (4x^3 + 3x^2 + 2x - 2) = 7$$

Example 6

Evaluate $\lim_{x \to 0} (x + 3)(3x - 3)(2x + 3)$

Solution

$$\lim_{x\to 0}(x + 3)(3x - 3)(2x + 3)$$

To evaluate the expression above, we will expand the function first

$$\lim_{x\to 0}(x + 3)(3x - 3)(2x + 3)$$

illustrate each equation for clarity;

$$(x + 3)(3x - 3)(2x + 3) = (A)(B)(C)$$

Multiply (A) and (B) first

$$(x + 3)(3x - 3) = x(3x - 3) + 3(3x - 3)$$
$$= (3x^2 - 3x) + (9x - 9)$$

Open the brackets

$$= 3x^2 - 3x + 9x - 9$$
$$= 3x^2 + 6x - 9$$

Therefore;

$$(A)(B) = 3x^2 + 6x - 9$$

Multiply $(3x^2 + 6x - 9)$ and (C)

$$(3x^2 + 6x - 9)(2x + 3)$$
$$(3x^2 + 6x - 9)(2x + 3) = 3x^2(2x + 3) + 6x(2x + 3) - 9(2x + 3)$$
$$= 6x^3 + 6x^2 + 12x^2 + 18x - 18x - 27$$
$$= 6x^3 + 18x^2 - 27$$

Having gotten the solution to the expression

$$(A)(B)(C) = 6x^3 + 18x^2 - 27$$
$$\lim_{x\to 0}(x + 3)(3x - 3)(2x + 3) = \lim_{x\to 0}(6x^3 + 18x^2 - 27)$$

Separate the functions using law 3

$$= \lim_{x\to 0}(6x^3 + 18x^2 - 27)$$

$$= \lim_{x \to 0} 6x^3 + \lim_{x \to 0} 18x^2 - \lim_{x \to 0} 27$$

Following law 2, separate the product of a constant and limit of a function

$$= 6\lim_{x \to 0} x^3 + 18\lim_{x \to 0} x^2 - \lim_{x \to 0} 27$$

To solve each limit, note that the limit of a constant is the constant itself,

$$6(0) + 18(0) - 27 = 0 + 0 - 27 = -27$$

Therefore,

$$\lim_{x \to 0}(x + 3)(3x - 3)(2x + 3) = -27$$

Exercise 2

Evaluate:

1. $\lim\limits_{x \to 0}(6x^5 - 5x^4 + 3x^3 + 5x^2 - x - 3 + a)$

2. Evaluate $\lim\limits_{x \to 0}(x^3 - 2x^2 + 4)$

3. Evaluate $\dfrac{\left(\lim\limits_{x \to 2} 3x^2 + x + 6\right)}{\left(\lim\limits_{x \to 1} 5x^2 - 2x + 8\right)}$

4. If $f_x = x^2 + 2$ and $g_x = 7x + 4$, find $\dfrac{\lim\limits_{x \to 2} f_x}{\lim\limits_{x \to 2} g_x}$

5. Evaluate $\lim_{x \to 1} \left(\dfrac{x^2 - 4x - 21}{x - 7} \right)$

6. Evaluate $\lim\limits_{x \to \alpha} \left(\dfrac{2x^3 + 2x^2 + 3x + 2}{5x^3 - 2x^2 + x - 5} \right)$

7. Evaluate $\lim_{x \to 3}(7x^4 - 3x^3 + 4x^2 - x - 1)$

8. Evaluate $\lim\limits_{x \to 2}(2x^3 - x^2 + 5)$

9. Given that;

$$P_1 = 2x^2 + 5x + 6, \qquad P_2 = 3x^2 - 2x + 1$$

find $\lim_{x \to 0}(P_1)(P_2)$

10.	Evaluate $\lim_{x \to 2}(x^2 + 2)(x + 3)$

Chapter 3

CONTINUITY OF A FUNCTION

A function is continuous at $x = a$ if the following conditions are met;

a) The function is defined at $x = a$. This means $f_{(a)}$ equals a real number. $f_{(a)}$ is defined.

b) The limit of the function exists as x approaches a. That means $\lim\limits_{x \to a} f_x$ exist.

c) The limit of the function as as x approaches a is equal to the function value at $x = a$.

that is; $\lim\limits_{x \to a} f_x = f_{(a)}$.

With the conditions stated above, you can know maybe a function is continuous or discontinuous.

Example 1
Determine whether f_x is continuous at at $x = 2$ and at $x = -1$.

$$f_x = \frac{x^2 - x - 2}{x + 1}$$

Solution

i.   How do we know f_x is continuous at $x = 2$?

First condition

$$f_x \ must \ be \ defined \ at \ at \ x = 2$$

$$f_2 = \frac{2^2 - 2 - 2}{2 + 1} = \frac{0}{3} = 0$$

$$f_2 = 0$$

Since $f_2 = 0$, the function f_2 **is** defined. The first condition is met.

Second condition

The limit of the function f_x must exist as x approaches 2.

$$\lim_{x \to a} f_x = \lim_{x \to 2} \frac{x^2 - x - 2}{x + 1}$$

$$\frac{2^2 - 2 - 2}{2 + 1} = 0$$

The limit, $\lim_{x \to 2} \frac{x^2 - x - 2}{x + 1}$ exist at 0. This made the second condition valid for

this function.

Third condition

The limit of the function as x approaches 2 must be equal to the function

value at $x = 2$.

That is;

$$\lim_{x \to 2} f_x = f_2$$

From the analysis $\lim_{x \to 2} f_x = 0$ *and* $f_2 = 0$. *This makes the condition valid.*

Since   the   three   conditions   were   met   for   the   function

$$f_x = \frac{x^2 - x - 2}{x + 1}$$

as x approaches **2.**

We can conclude that f_x is continuous at $x = 2$.

ii.   How do we know f_x is continuous at $x = -1$.

First condition

$$f_x \text{ must be defined at at } x = -1$$

$$f_{-1} = \frac{-1^2 - (-1) - 2}{-1 + 1} = \frac{0}{0}$$

$$f_{-1} = \frac{0}{0}$$

Since $f_{-1} = \frac{0}{0}$, the function f_{-1} **is not** defined. The first condition is not met.

Since the first condition is not met. The function f_x is discontinuous at $x = -1$.

Example 2

Determine whether f_x is continuous at at $x = 1$. $f_x = \frac{x^3-1}{x-1}$

Solution

How do we know f_x is continuous at $x = 1$?

First condition

$$f_x \; must \; be \; defined \; at \; at \; x = 1$$

$$f_1 = \frac{1^3 - 1}{1 - 1} = \frac{0}{0}$$

$$f_{-1} = \frac{0}{0}$$

Since $f_1 = \frac{0}{0}$, the function f_1 **is not** defined. The first condition is not met.

Since the first condition is not met. The function f_x is discontinuous at $x = 1$.

Example 3

Determine whether f_x is continuous at at $x = 2$ and at $x = -1$.

$$f_x = \begin{vmatrix} \frac{x^3-1}{x-1} & x \neq 1 \\ 2 & x = 1 \end{vmatrix}$$

Solution

i.   How do we know f_x is continuous at $x = 1$?

First condition

$$f_x \text{ must be defined at at } x = 1$$
$$f_1 = 2$$

Since $f_1 = 2$, the function f_1 **is** defined. The first condition is met.

Second condition

The limit of the function f_x must exist as x approaches 1.

$$\lim_{x \to a} f_{x=} \lim_{x \to 1}\frac{x^3-1}{x-1}$$

Note

$$a^3 - b^3 = (a - b)(a^2 + ab + b^2)$$

Therefore:

$$x^3 - 1 = x^3 - 1^3$$
$$x^3 - 1^3 = (x - 1)(x^2 + x + 1)$$
$$\lim_{x \to 1}\frac{x^3 - 1}{x - 1} = \lim_{x \to 1}\frac{(x - 1)(x^2 + x + 1)}{x - 1}$$
$$\lim_{x \to 1} x^2 + x + 1 = 1 + 1 + 1 = 3$$
$$\lim_{x \to 1}\frac{x^3 - 1}{x - 1} = 3$$

The limit $\lim_{x \to 1}\frac{x^3-1}{x-1}$ exist at 3. This made the second condition valid for this function.

Third condition

The limit of function as x approaches 1 must be equal to the function value at $x = 1$.

That is;

$$\lim_{x \to 1} f_x = f_1$$

From the analysis $\lim_{x \to 1} f_x = 3$ and $f_1 = 2$. This condition is not valid as;

$$\lim_{x \to 1} f_x \neq f_1$$

Since the third condition is not met.

The function f_x is discontinuous at $x = 1$.

Exercise 3: Try this on your own

1. Examine $f_x = 7x^2 + 3x + 8$ for continuity at $x = -1$.

2. Examine $h_x = 4x^3 + 3x^2 + 2x - 1$ for continuity at $x = 2$.

Chapter 4

THE GRADIENT FUNCTION

Gradient Of A Straight Line

The gradient of a straight line is the ratio of increase in y to *increase in x* in moving from one point to another.

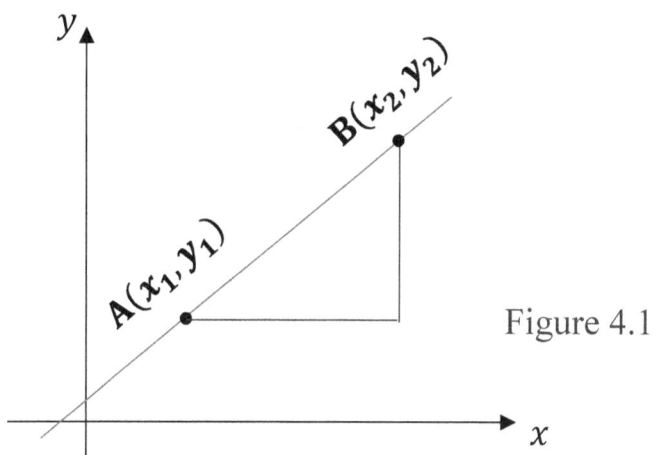

Figure 4.1

From figure 4.1 above, if $A(x_1, y_1)$ and $B(x_2, y_2)$ are two points on the line, the gradient of the straight line is given as;

$$m = \frac{y_2 - y_1}{x_2 - x_1}$$

The gradient (m) will be constant along the line and equal to the tangent of the angle θ made with the positive direction of the $x - axis$.

Gradient Function

The gradient function gives the slope of the gradient of a function at any point on its curve.

If an equation is given by $y = f_{(x)}$ let's consider the curve below.

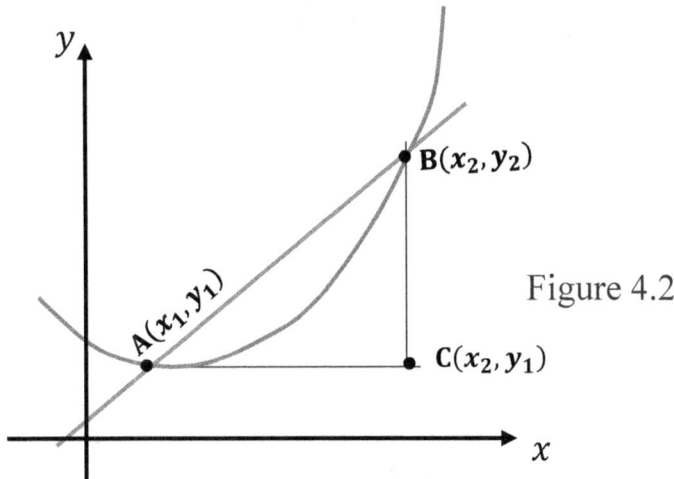

Figure 4.2

From figure 4.2, we have a secant line AB (A secant line is a line that intercepts the curve at a minimum of two distinct points.

Let m be the gradient of the secant AB, then

$$m = \frac{increase\ in\ y}{increase\ in\ x}$$

$$m = \frac{y_2 - y_1}{x_2 - x_1}$$

Where;

$$\Delta x = x_2 - x_1, and\ \Delta y = y_2 - y_1$$

Δx reads "delta x" and Δy reads "delta y".

Let point A in figure 4.2 be fixed as point B move closer to A along the curve. You will notice that the closer point B to point A, along the curve, the closer Δx approaches 0. Therefore, the slope or gradient of secant AB approaches the slope of the tangent at A.

If Δx is made to approach 0, the function;

$$m_x = \frac{f_{(x+\Delta x)} - f_{(x)}}{\Delta x}$$

Is called the gradient function of $y = f_{(x)}$.

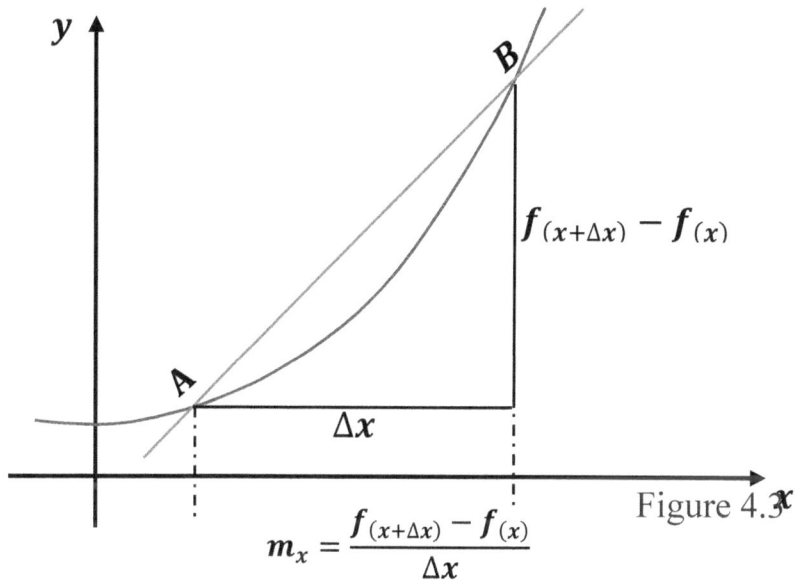

$$m_x = \frac{f_{(x+\Delta x)} - f_{(x)}}{\Delta x}$$

Figure 4.3

DERIVATIVE OF A FUNCTION

A derivative is a function that measures the slope or gradient. It is dependent on x, and it is found by differentiating a function of the form $y = f_{(x)}$. It measures the change of the function value, which is the output value to the change of its input value.

The process of finding a derivative is called Differentiation.

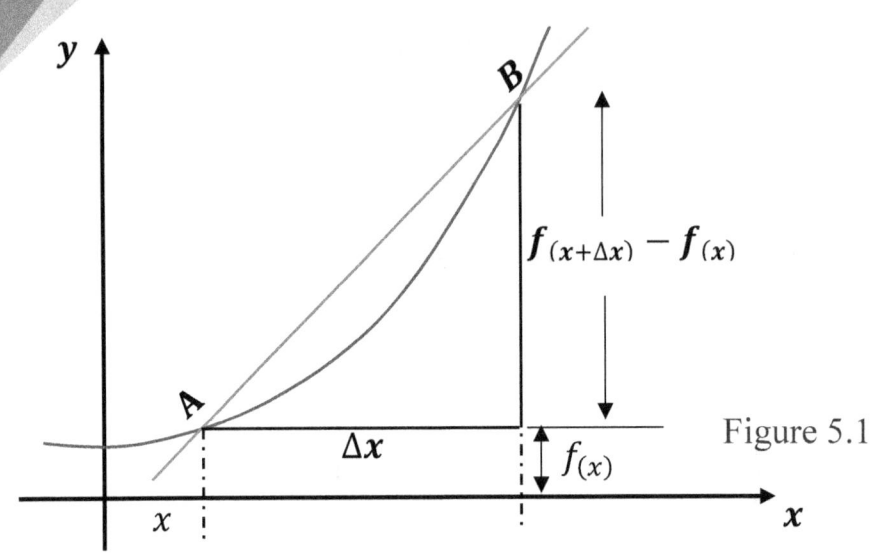

Figure 5.1

Secant Line

A secant line is a straight line joining two points on a curve of a function. It is equivalent to the slope between two points on a curve. It is the average rate of change.

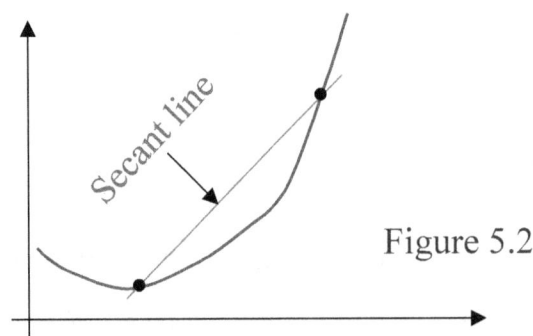

Figure 5.2

Tangent Line

A tangent line is a straight line that touches a curve at only one point. The gradient or slope of the tangent line at the point on the curve is known as the derivative of that point. It shows the instantaneous rate of change of the function at that one point.

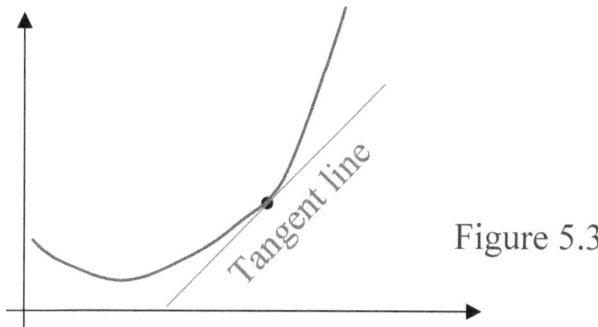

Figure 5.3

From figure 5.1, recall that, gradient function of a curve $y = f_{(x)}$ is;

$$m_1 = \frac{f_{(x+\Delta x)} - f_{(x)}}{\Delta x}$$

This expression m_1 represent the gradient of the secant AB. As $point\ B$ moves closer to point $point\ A$, Δx becomes smaller, and it tends to 0. As B clashes on point A, the gradient becomes the gradient of a tangent at point A.

Let gradient of the tangent at point A be m_2.

In simple words, as B approaches A, along the curve, gradient m_1 tends to gradient m_2 or the gradient of secant AB tends to the gradient of the tangent to the curve at A.

As Δx tends to 0 ($\Delta x \to 0$), m_2 becomes the limiting value of m_1.

The limiting value is represented as;

$$f'_{(x)} = \lim_{\Delta x \to 0} \frac{f_{(x+\Delta x)} - f_{(x)}}{\Delta x}$$

It is denoted as $f'_{(x)}$ and it is called the **derivative of $f_{(x)}$**.

Summarily;

For better understanding, figure 5.1 will be redrawn to reflect function y and its increment as x increases. This clearly explains the first principle of Differentiation.

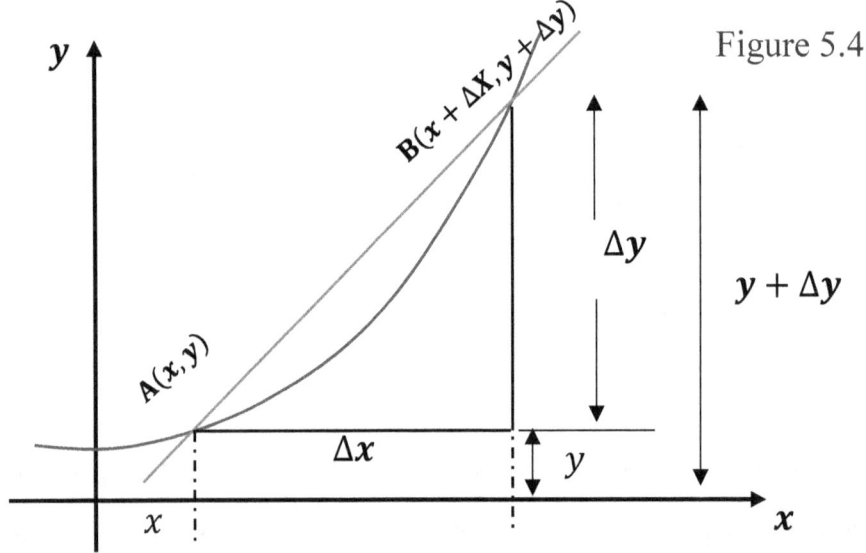

Figure 5.4

for $y = f_{(x)}$, **a** small change or increment in x will cause a change or increment in y.

Δx is the increment in x

Δy is the increment in y

Therefore

$$y = f_{(x)}$$
$$y + \Delta y = f_{(x+\Delta x)}$$
$$\Delta y = f_{(x+\Delta x)} - y$$

Where;

$$y = f_{(x)}$$
$$\Delta y = f_{(x+\Delta x)} - f_{(x)}$$

If m_1 is the gradient function of the secant AB.

$$m_1 = \frac{y_2 - y_1}{x_2 - x_1} = \frac{\Delta y}{\Delta x}$$

Therefore;

$$m_1 = \frac{\Delta y}{\Delta x} = \frac{f_{(x+\Delta x)} - f_{(x)}}{\Delta x}$$

As B approaches A, the gradient of the tangent at A is m_2

Then;

m_2 *is the limiting value of* m_1 *as* $\Delta x \to 0$.

Mathematically;

$$m_2 = \lim_{\Delta x \to 0} \frac{\Delta y}{\Delta x} = \frac{dy}{dx}$$

$\frac{dy}{dx}$ *reads dee y dee x*. It is the limiting value of the slope or gradient of secant AB as point B get closer to point A. it is also known as **Differentiation;** *which is the process of finding the derivative of a function.*

Notation Used For Derivative Of A Function

$f'(x) \to f\ prime\ of\ x$

$\frac{d}{dx} f \to dee - dee\ x\ of\ f$

$\frac{dy}{dx} \to dee - y\ dee - x$

Most commonly used are $f'_{(x)}$ and $\frac{dy}{dx}$.

Chapter 5

DIFFERENTIAL CALCULUS

Differentiation is the process of finding the gradient function $\frac{dy}{dx}$ of f_x', which is the derivative of $y\ with\ respect\ to$ x. it is also called the differential coefficient of y. The gradient at a point on the curve given by $y = f_{(x)}.$is $\frac{dy}{dx}$.

$\frac{dy}{dx}$ gives the actual gradient function. It measures the rate of change of the quantity y as compared with x. It shows the instantaneous rate of change. For example, if $y = x^2$, $\frac{dy}{dx} = 2x$, then the rate of change of y as compared with x is $2x$.

It is important to understand that $\frac{dy}{dx}$ measures the rate of change of y as x. It was this idea that made calculus one of the great inventions of mathematics. The notation and the idea can be generally applied. For example, if $s = f_{(x)}$, then $\frac{ds}{dt}$ is the derivative of s with respect to x or measures the rate of change of $s\ wrt\ t$.

Also, $A = r^2$, then $\frac{dA}{dr} = 2r$. A is changing 2r times as much as r is changing.

Differentiation From First Principle

As explained in the previous chapter, the process involved in finding the derivative of a function using the limiting value is called Differentiation. From the first principle

Example 1
Find the derivative of $f_{(x)} = x^2$, using the first principle.
Solution

$$f_{(x)} = x^2$$

Apply increment

$$f_{(x+\Delta x)} = (x + \Delta x)^2$$

Expand the equation

$$f_{(x+\Delta x)} = (x + \Delta x)(x + \Delta x)$$

$$f_{(x+\Delta x)} = x^2 + 2x\Delta x + (\Delta x)^2$$

Subtract $f_{(x)}$ from both sides

$$f_{(x+\Delta x)} - f_{(x)} = x^2 + 2x\Delta x + (\Delta x)^2 - f_{(x)}$$

$$f_{(x+\Delta x)} - f_{(x)} = x^2 + 2x\Delta x + (\Delta x)^2 - x^2$$

$$f_{(x+\Delta x)} - f_{(x)} = 2x\Delta x + (\Delta x)^2$$

Finding the gradient by dividing both sides by Δx

$$\frac{f_{(x+\Delta x)} - f_{(x)}}{\Delta x} = \frac{2x\Delta x + (\Delta x)^2}{\Delta x}$$

$$\frac{f_{(x+\Delta x)} - f_{(x)}}{\Delta x} = \frac{2x\Delta x}{\Delta x} + \frac{(\Delta x)^2}{\Delta x}$$

$$\frac{f_{(x+\Delta x)} - f_{(x)}}{\Delta x} = 2x + \Delta x$$

Take $\lim\limits_{\Delta x \to 0} \frac{\Delta y}{\Delta x} = f'_x$

$$\lim\limits_{\Delta x \to 0} \frac{f_{(x+\Delta x)} - f_{(x)}}{\Delta x} = \lim\limits_{\Delta x \to 0} 2x + \Delta x = 2x$$

$$f'_x = 2x$$

$$\frac{dy}{dx} = f'_x = 2x$$

Example 2
Find the derivative of $y = x^3$, using the first principle.

Solution

$$y = x^3$$

Apply increment

$$y + \Delta y = (x + \Delta x)^3$$

Expand the equation

$$(x + \Delta x)^3 = (x + \Delta x)(x + \Delta x)(x + \Delta x)$$
$$= (x^2 + 2x\Delta x + (\Delta x)^2)(x + \Delta x)$$
$$= x^3 + x^2\Delta x + 2x^2\Delta x + 2x(\Delta x)^2 + x(\Delta x)^2 + (\Delta x)^3$$
$$(x + \Delta x)^3 = x^3 + 3x^2\Delta x + 3x(\Delta x)^2 + (\Delta x)^3$$
$$y + \Delta y = (x + \Delta x)^3$$
$$y + \Delta y = x^3 + 3x^2\Delta x + 3x(\Delta x)^2 + (\Delta x)^3$$

Subtract y from both sides

$$y + \Delta y - y = x^3 + 3x^2\Delta x + 3x(\Delta x)^2 + (\Delta x)^3 - y$$
$$\Delta y = x^3 + 3x^2\Delta x + 3x(\Delta x)^2 + (\Delta x)^3 - x^3$$
$$\Delta y = 3x^2\Delta x + 3x(\Delta x)^2 + (\Delta x)^3$$

Finding the gradient by dividing both sides by Δx

$$\frac{\Delta y}{\Delta x} = \frac{3x^2\Delta x}{\Delta x} + \frac{3x(\Delta x)^2}{\Delta x} + \frac{(\Delta x)^3}{\Delta x}$$
$$\frac{\Delta y}{\Delta x} = 3x^2 + 3x\Delta x + (\Delta x)^2$$

Take $\lim\limits_{\Delta x \to 0} \dfrac{\Delta y}{\Delta x} = \dfrac{dy}{dx}$

$$\lim\limits_{\Delta x \to 0} \frac{\Delta y}{\Delta x} = \lim\limits_{\Delta x \to 0} 3x^2 + 3x\Delta x + (\Delta x)^2 = 3x^2$$
$$\frac{dy}{dx} = 3x^2$$

Example 3

Find the derivative of $y = ax + b$, using the first principle. Where a and b are constants.

Solution

$$y = ax + b$$

Apply increment

$$y + \Delta y = a(x + \Delta x) + b$$

Expand the equation

$$y + \Delta y = ax + a\Delta x + b$$

Subtract y from both sides

$$y + \Delta y - y = ax + a\Delta x + b - y$$

Where $y = ax + b$

$$\Delta y = ax + a\Delta x + b - (ax + b)$$

$$\Delta y = ax + a\Delta x + b - ax - b$$

$$\Delta y = ax - ax + a\Delta x + b - b$$

$$\Delta y = a\Delta x$$

Finding the gradient by dividing both sides by Δx

$$\frac{\Delta y}{\Delta x} = \frac{a\Delta x}{\Delta x}$$

$$\frac{\Delta y}{\Delta x} = a$$

Take $\lim\limits_{\Delta x \to 0} \frac{\Delta y}{\Delta x} = \frac{dy}{dx}$

$$\lim\limits_{\Delta x \to 0} \frac{\Delta y}{\Delta x} = \lim\limits_{\Delta x \to 0} a$$

$$\frac{dy}{dx} = a$$

Example 4
Find the derivative of $y = 3x^2 - x + 5$, using the first principle.

Solution

$$y = 3x^2 - x + 5$$

Apply increment

$$y + \Delta y = 3(x + \Delta x)^2 - (x + \Delta x) + 5$$

Expand the equation

$$y + \Delta y = 3(x^2 + 2x\Delta x + (\Delta x)^2) - (x + \Delta x) + 5$$

$$y + \Delta y = 3x^2 + 6x\Delta x + 3(\Delta x)^2 - x - \Delta x + 5$$

$$y + \Delta y = 3x^2 + 6x\Delta x - \Delta x + 3(\Delta x)^2 - x + 5$$

$$y + \Delta y = 3x^2 + 5x\Delta x + 3(\Delta x)^2 - x + 5$$

Subtract y from both sides

$$y + \Delta y - y = 3x^2 + 5x\Delta x + 3(\Delta x)^2 - x + 5 - y$$

Where $y = 3x^2 - x + 5$

$$\Delta y = 3x^2 + 5x\Delta x + 3(\Delta x)^2 - x + 5 - (3x^2 - x + 5)$$

$$\Delta y = 3x^2 + 5x\Delta x + 3(\Delta x)^2 - x + 5 - 3x^2 + x - 5$$

$$\Delta y = 3x^2 - 3x^2 + 5x\Delta x + 3(\Delta x)^2 - x + x + 5 - 5$$

$$\Delta y = 5x\Delta x + 3(\Delta x)^2$$

Finding the gradient by dividing both sides by Δx

$$\frac{\Delta y}{\Delta x} = \frac{5x\Delta x + 3(\Delta x)^2}{\Delta x}$$

$$\frac{\Delta y}{\Delta x} = \frac{5x\Delta x}{\Delta x} + \frac{3(\Delta x)^2}{\Delta x}$$

$$\frac{\Delta y}{\Delta x} = 5x + 3\Delta x$$

Take $\lim\limits_{\Delta x \to 0} \frac{\Delta y}{\Delta x} = \frac{dy}{dx}$

$$\lim_{\Delta x \to 0} \frac{\Delta y}{\Delta x} = \lim_{\Delta x \to 0} 5x + 3\Delta x = 5x$$

$$\frac{dy}{dx} = 5x$$

Formular method of finding the derivative of a function.

There may be some complex functions that will require a lot of pages to find their derivative. The use of the formula method will simplify the whole process.

Derivative Formula Of Function Of y Without Coefficient

If $y = x^n$

find the derivative $(\frac{dy}{dx})$ of this function using the first principal method.

$$y = x^n$$

An increment in x will cause an increment in y.

$$y + \Delta y = (x + \Delta x)^n$$

Using the binomial expansion principle, let's expand

If

$$(a + b)^n = C_0^n a^n + C_1^n a^{n-1} b + C_2^n a^{n-2} b^2 - - - + C_r^n a^{n-r} b^r - - - + b^n$$

Where

$$C_r^n = \frac{n!}{(n - r)! \, r!}$$

Therefore

$$(x + \Delta x)^n = C_0^n x^n + C_1^n x^{n-1} \Delta x + C_2^n x^{n-2} (\Delta x)^2 - - - + C_r^n x^{n-r} (\Delta x)^r - - - - + (\Delta x)^n$$

Analysis

$$C_0^n x^n = \frac{n!}{(n-0)!\,0!} \times x^n = x^n \times 1 = x^n, where\ r = 0$$

$$C_1^n x^{n-1} \Delta x = \frac{n!}{(n-1)!\,1!} \times x^{n-1} \times \Delta x =$$

$$= \frac{n(n-1)!}{(n-1)!\,1!} \times x^{n-1} \times \Delta x$$

$$= nx^{n-1} \times \Delta x = nx^{n-1}\Delta x, where\ r = 1$$

$$C_2^n x^{n-2} (\Delta x)^2 = \frac{n!}{(n-2)!\,2!} \times x^{n-2} \times (\Delta x)^2 =$$

$$= \frac{n(n-1)(n-2)!}{(n-2)!\,2!} \times x^{n-2} \times (\Delta x)^2$$

$$= \frac{n(n-1)}{2} x^{n-2} \times (\Delta x)^2 = \frac{n(n-1)}{2} x^{n-2}(\Delta x)^2, where\ r = 2$$

$$(x + \Delta x)^n = x^n + nx^{n-1}\Delta x + \frac{n(n-1)}{2} x^{n-2}(\Delta x)^2 - - - + (\Delta x)^n$$

$$y + \Delta y = x^n + nx^{n-1}\Delta x + \frac{n(n-1)}{2} x^{n-2}(\Delta x)^2 - - - + (\Delta x)^n$$

Subtract y from both sides

$$\Delta y = x^n + nx^{n-1}\Delta x + \frac{n(n-1)}{2} x^{n-2}(\Delta x)^2 - - - + (\Delta x)^n - y$$

Recall

$$y = x^n$$

$$\Delta y = x^n + nx^{n-1}\Delta x + \frac{n(n-1)}{2} x^{n-2}(\Delta x)^2 - - - + (\Delta x)^n - x^n$$

$$\Delta y = nx^{n-1}\Delta x + \frac{n(n-1)}{2} x^{n-2}(\Delta x)^2 - - - + (\Delta x)^n$$

Divide both sides by Δx

$$\frac{\Delta y}{\Delta x} = \frac{nx^{n-1}\Delta x + \frac{n(n-1)}{2}x^{n-2}(\Delta x)^2 - - - + (\Delta x)^n}{\Delta x}$$

$$\frac{\Delta y}{\Delta x} = \frac{nx^{n-1}\Delta x}{\Delta x} + \frac{n(n-1)}{2\,\Delta x}x^{n-2}(\Delta x)^2 - - - + \frac{(\Delta x)^n}{\Delta x}$$

$$\frac{\Delta y}{\Delta x} = nx^{n-1} + \frac{n(n-1)}{2}x^{n-2}\Delta x - - - + \frac{(\Delta x)^n}{\Delta x}$$

Take $\lim\limits_{\Delta x \to 0} \frac{\Delta y}{\Delta x}$

$$\lim_{\Delta x \to 0}\frac{\Delta y}{\Delta x} = \lim_{\Delta x \to 0} nx^{n-1} + \frac{n(n-1)}{2}x^{n-2}\Delta x - - - + \frac{(\Delta x)^n}{\Delta x}$$

$$\frac{dy}{dx} = nx^{n-1}$$

Hence;

For $y = x^n$

$$\frac{dy}{dx} = nx^{n-1}$$

This is true for the integral and fractional value of n.

Examples

Find the derivative of the following using formular method.
  a. x^6
  b. $x^{\frac{1}{4}}$
  c. $\sqrt[4]{x^3}$

Solution

  a. If $y = x^6$

For $y = x^n$, $\frac{dy}{dx} = nx^{n-1}$

$y = x^6$, where $n = 6$

Therefore

$$\frac{dy}{dx} = 6 \times x^{6-1} = 6x^5$$

b. If $y = x^{\frac{1}{4}}$

For $y = x^n$, $\frac{dy}{dx} = nx^{n-1}$

$y = x^{\frac{1}{4}}$, where $n = \frac{1}{4}$

Therefore

$$\frac{dy}{dx} = \frac{1}{4} \times x^{\frac{1}{4}-1} = \frac{1}{4}x^{\frac{-3}{4}} = \frac{1}{4\sqrt[4]{x^3}}$$

c. If $y = \sqrt[4]{x^3}$

For $y = x^n$, $\frac{dy}{dx} = nx^{n-1}$

$y = \sqrt[4]{x^3} = x^{\frac{3}{4}}$, where $n = \frac{3}{4}$

Therefore

$$\frac{dy}{dx} = \frac{3}{4} \times x^{\frac{3}{4}-1} = \frac{3}{4}x^{\frac{-1}{4}} = \frac{3}{4\sqrt[4]{x}}$$

The derivative formula of a function of y with coefficient

When you have a function of a form

$$y = af_x$$

Then an increment in x will cause an increment in y

$$y + \Delta y = af_{(x+\Delta x)}$$

Subtract y from both sides

$$y + \Delta y - y = af_{(x+\Delta x)} - y$$
$$\Delta y = af_{(x+\Delta x)} - y$$

Where

$$y = af_x$$

Therefore

$$\Delta y = af_{(x+\Delta x)} - af_x$$

Find the gradient by dividing both sides by Δx

$$\frac{\Delta y}{\Delta x} = \frac{af_{(x+\Delta x)} - af_x}{\Delta x}$$

$$\frac{\Delta y}{\Delta x} = a\left(\frac{f_{(x+\Delta x)} - f_x}{\Delta x}\right)$$

Take $\lim\limits_{\Delta x \to 0} \frac{\Delta y}{\Delta x}$

$$\lim\limits_{\Delta x \to 0} \frac{\Delta y}{\Delta x} = \lim\limits_{\Delta x \to 0} a\left(\frac{f_{(x+\Delta x)} - f_x}{\Delta x}\right)$$

Therefore

$$\frac{dy}{dx} = af'_x$$

If $f_x = x^n$

$$f'_x = nx^{n-1}$$

$$\frac{dy}{dx} = anx^{n-1}$$

For a function with coefficient of a

$$y = ax^n$$

$$\frac{dy}{dx} = anx^{n-1}$$

This is true for the integral and fractional value of n.

Note: the derivative of a constant is 0

If $y = c$

Where c is a constant

then $\frac{dy}{dx} = 0$

Examples

Find the derivative of the following using formular method.

    a. $y = -2x^3$

    b. $y = \frac{-2}{3}x^4$

    c. $y = 3x^{\frac{1}{2}}$

Solution

    a. If $y = -2x^3$

        For $y = ax^n$, $\frac{dy}{dx} = anx^{n-1}$

       $y = -2x^3$, where $n = 3$

       Therefore

$$\frac{dy}{dx} = -2 \times 3 \times x^{3-1} = -6x^2$$

    b. If $y = \frac{-2}{3}x^4$

        For $y = ax^n$, $\frac{dy}{dx} = anx^{n-1}$

$y = \dfrac{-2}{3}x^4$, where $n = 4$

Therefore

$$\frac{dy}{dx} = \frac{-2}{3} \times 4 \times x^{4-1} = \frac{-8}{3}x^3$$

c. If $y = 3x^{\frac{1}{2}}$

For $y = ax^n$, $\dfrac{dy}{dx} = anx^{n-1}$

$y = 3x^{\frac{1}{2}}$, where $n = \dfrac{1}{2}$

Therefore

$$\frac{dy}{dx} = 3 \times \frac{1}{2} \times x^{\frac{1}{2}-1} = \frac{3}{2}x^{\frac{-1}{2}} = \frac{3}{2\sqrt{x}}$$

Derivative Of Sum Of Function

If $f, u, and \ v \ are \ function \ of \ x$, such that

$$f_{(x)} = u_{(x)} + v_{(x)}$$

$$f_{(x+\Delta x)} = u_{(x+\Delta x)} + v_{(x+\Delta x)}$$

$$f_{(x+\Delta x)} - f_{(x)} = u_{(x+\Delta x)} + v_{(x+\Delta x)} - u_{(x)} - v_{(x)}$$

$$f_{(x+\Delta x)} - f_{(x)} = u_{(x+\Delta x)} - u_{(x)} + v_{(x+\Delta x)} - v_{(x)}$$

The gradient of the functions

$$\frac{f_{(x+\Delta x)} - f_{(x)}}{\Delta x} = \frac{u_{(x+\Delta x)} - u_{(x)}}{\Delta x} + \frac{v_{(x+\Delta x)} - v_{(x)}}{\Delta x}$$

Take $\lim\limits_{\Delta x \to 0} \dfrac{\Delta y}{\Delta x}$

$$\lim_{\Delta x \to 0} \frac{f_{(x+\Delta x)} - f_{(x)}}{\Delta x} = \lim_{\Delta x \to 0} \frac{u_{(x+\Delta x)} - u_{(x)}}{\Delta x} + \frac{v_{(x+\Delta x)} - v_{(x)}}{\Delta x}$$

$$f'_x = u'_x + v'_x$$

$$\frac{dy}{dx} = \frac{du}{dx} + \frac{dv}{dx}$$

Examples 1

Differentiate with respect to x; $2x^3 - 5x^2 + 2$

Solution

$$y = 2x^3 - 5x^2 + 2$$

For $y = ax^n$, $\frac{dy}{dx} = anx^{n-1}$

$$\frac{dy}{dx} = 3 \times 2x^{3-1} - 2 \times 5x^{2-1} + 0$$

$$\frac{dy}{dx} = 6x^2 - 10x^1 = 6x^2 - 10x$$

Examples 2

Differentiate with respect to x; $3x^2 + \frac{1}{x}$

Solution

$$y = 3x^2 + \frac{1}{x} = 3x^2 + x^{-1}$$

For $y = ax^n$, $\frac{dy}{dx} = anx^{n-1}$

$$\frac{dy}{dx} = 2 \times 3x^{2-1} + (-1)x^{-1-1}$$

$$\frac{dy}{dx} = 6x - x^{-2} = 6x - \frac{1}{x^2}$$

Examples 3

Differentiate with respect to x; $\frac{x^3+2x^2+1}{x}$

Solution

$$y = \frac{x^3 + 2x^2 + 1}{x} = \frac{x^3}{x} + \frac{2x^2}{x} + \frac{1}{x}$$

$$y = x^2 + 2x + x^{-1}$$

For $y = ax^n$, $\frac{dy}{dx} = anx^{n-1}$

$$\frac{dy}{dx} = 2 \times x^{2-1} + 2x^{1-1} + x^{-1-1}$$

$$\frac{dy}{dx} = 2x^1 + 2x^0 + x^{-2} = 2x + 2 + \frac{1}{x^2}$$

Examples 4

Differentiate with respect to x; $\sqrt{x} + \frac{1}{\sqrt{x}} - 3$

Solution

$$y = \sqrt{x} + \frac{1}{\sqrt{x}} - 3 = x^{\frac{1}{2}} + x^{-\frac{1}{2}} - 3$$

$$y = x^{\frac{1}{2}} + x^{-\frac{1}{2}} - 3$$

For $y = ax^n$, $\frac{dy}{dx} = anx^{n-1}$

$$\frac{dy}{dx} = \frac{1}{2} \times x^{\frac{1}{2}-1} + \left(-\frac{1}{2}\right)x^{-\frac{1}{2}-1} - 0$$

$$\frac{dy}{dx} = \frac{1}{2} \times x^{-\frac{1}{2}} + \left(-\frac{1}{2}\right)x^{-\frac{3}{2}}$$

$$\frac{dy}{dx} = \frac{1}{2} \times x^{-\frac{1}{2}} - \frac{1}{2}x^{-\frac{3}{2}}$$

$$\frac{dy}{dx} = \frac{1}{2\sqrt{x}} - \frac{1}{2\sqrt{x^3}}$$

Exercise 5

Evaluate:

1. Find from first principle the derivative of $y = \frac{3}{x}$ *with respect to x*.

2. Find from first principle the derivative of $y = 2x^2 - x$ with respect to x.

3. Find from first principle the derivative of $y = x - \dfrac{1}{x}$ with respect to x.

4. Find from first principle the derivative of $y = 5x^2 - 2$ with respect to x.

5. Find from first principle the derivative of $y = 3x^2$ with respect to x.

6. Using the formula method, find the derivative of $y = 4x^3 - 3x^2 - 5x + 3$ with respect to x

7. Using the formula method, find the derivative of $y = x^7$ with respect to x

8. Using the formula method, find the derivative of $y = 7x^5$ with respect to x.

9. Using the formula method, find the derivative of $y = \dfrac{9}{x^2}$ with respect to x

10. Using the formula method, find the derivative of $y = \sqrt{x} + \dfrac{1}{\sqrt{x}}$ with respect to x

Chapter 6

TECHNIQUES IN DIFFERENTIATION

There are different methods used for finding derivatives. Here in this chapter, we shall thoroughly consider all the methods.

Function Of A Function

It is also called composite function or chain rule for differentiation. It is one type of method used for finding the derivative of a function.

In this case;

$$y \text{ is a function of } u, \text{ and } u \text{ is a function of } x$$

$$y = f_{(u)}$$

$$u = h_{(x)}$$

$$\frac{\Delta y}{\Delta x} = \frac{\Delta y}{\Delta u} \times \frac{\Delta u}{\Delta x}$$

Take $\lim\limits_{\Delta x \to 0} \dfrac{\Delta y}{\Delta x} = \dfrac{dy}{dx}$

$$\lim_{\Delta x \to 0} \frac{\Delta y}{\Delta x} = \lim_{\Delta x \to 0} \left(\frac{\Delta y}{\Delta u} \times \frac{\Delta u}{\Delta x} \right)$$

$$\lim_{\Delta x \to 0} \frac{\Delta y}{\Delta x} = \lim_{\Delta x \to 0} \frac{\Delta y}{\Delta u} \times \lim_{\Delta x \to 0} \frac{\Delta u}{\Delta x}$$

As $\Delta x \to 0, \Delta u \to 0$

$$\lim_{\Delta x \to 0} \frac{\Delta y}{\Delta u} = \lim_{\Delta u \to 0} \frac{\Delta y}{\Delta u}$$

$$\lim_{\Delta x \to 0} \frac{\Delta y}{\Delta x} = \lim_{\Delta u \to 0} \frac{\Delta y}{\Delta u} \times \lim_{\Delta x \to 0} \frac{\Delta u}{\Delta x}$$

$$\frac{dy}{dx} = \frac{dy}{du} \times \frac{du}{dx}$$

Examples 1

Differentiate with respect to x; $y = \frac{1}{(3x-4)^2}$

<u>Solution</u>

$$y = \frac{1}{(3x - 4)^2}$$

Let $u = (3x - 4)$

Then $y = \frac{1}{(u)^2}$

For $y = ax^n$, $\frac{dy}{dx} = anx^{n-1}$

Differentiate the function u and y.

$$u = 3x - 4, \quad \frac{du}{dx} = 3$$

$$y = \frac{1}{(u)^2} = u^{-2}, \frac{dy}{du} = -2u^{-2-1} = -2u^{-3}$$

$$\frac{dy}{dx} = \frac{dy}{du} \times \frac{du}{dx}$$

$$\frac{dy}{dx} = -2u^{-3} \times 3 = -6u^{-3}$$

Recall $u = 3x - 4$

$$\frac{dy}{dx} = -6u^{-3} = -6(3x - 4)^{-3}$$

$$\frac{dy}{dx} = -6(3x - 4)^{-3}$$

Examples 2
Differentiate with respect to x; $y = \sqrt{2x + 5}$

<u>Solution</u>

$$y = \sqrt{2x + 5}$$

Let $u = (2x + 5)$

Then $y = \sqrt{u}$

For $y = ax^n$, $\frac{dy}{dx} = anx^{n-1}$

Differentiate the function u *and* y.

$$u = 2x + 5,$$

$$\frac{du}{dx} = 2$$

$$y = \sqrt{u} = u^{\frac{1}{2}},$$

$$\frac{dy}{du} = \frac{1}{2}u^{\frac{1}{2}-1} = \frac{1}{2}u^{-\frac{1}{2}} = \frac{1}{2\sqrt{u}}$$

$$\frac{dy}{dx} = \frac{dy}{du} \times \frac{du}{dx}$$

$$\frac{dy}{dx} = \frac{1}{2\sqrt{u}} \times 2 = \frac{1}{\sqrt{u}}$$

Recall $u = 2x + 5$

$$\frac{dy}{dx} = \frac{1}{\sqrt{u}} = \frac{1}{\sqrt{2x + 5}}$$

$$\frac{dy}{dx} = \frac{1}{\sqrt{2x + 5}}$$

Examples 3

Differentiate with respect to x; $y = (3x^2 - 2)^3$

<u>Solution</u>

$$y = (3x^2 - 2)^3$$

Let $u = 3x^2 - 2$

Then $y = u^3$

$$\text{For } y = ax^n, \quad \frac{dy}{dx} = anx^{n-1}$$

Differentiate the function u **and** y.

$$u = 3x^2 - 2$$

$$\frac{du}{dx} = 6x$$

$$y = u^3,$$

$$\frac{dy}{du} = 3u^2$$

$$\frac{dy}{dx} = \frac{dy}{du} \times \frac{du}{dx}$$

$$\frac{dy}{dx} = 3u^2 \times 6x = 18xu^2$$

Recall $u = 3x^2 - 2$

$$\frac{dy}{dx} = 18xu^2 = 18x(3x^2 - 2)^2$$

$$\frac{dy}{dx} = 18x(3x^2 - 2)^2$$

Examples 4

Differentiate with respect to x; $y = \dfrac{5}{(4-x^2)^3}$

Solution

$$y = \frac{5}{(4 - x^2)^3}$$

Let $u = (4 - x^2)$

Then $y = \dfrac{5}{(u)^3}$

For $y = ax^n$, $\frac{dy}{dx} = anx^{n-1}$

Differentiate the function u and y.

$$u = 4 - x^2$$

$$\frac{du}{dx} = -2x$$

$$y = \frac{5}{(u)^3} = 5u^{-3}$$

$$\frac{dy}{du} = 5 \times (-3)u^{-3-1} = -15u^{-4}$$

$$\frac{dy}{dx} = \frac{dy}{du} \times \frac{du}{dx}$$

$$\frac{dy}{dx} = -15u^{-4} \times (-2x) = 30xu^{-4}$$

Recall $u = 4 - x^2$

$$\frac{dy}{dx} = 30xu^{-4} = 30x(4 - x^2)^{-4}$$

$$\frac{dy}{dx} = \frac{30x}{30x(4 - x^2)^4}$$

Examples 5

Differentiate with respect to x; $y = \dfrac{2}{(3x^2 - x + 7)^3}$

Solution

$$y = \frac{2}{(3x^2 - x + 7)^3}$$

Let $u = 3x^2 - x + 7$

Then $y = \dfrac{2}{(u)^3}$

For $y = ax^n$, $\frac{dy}{dx} = anx^{n-1}$

Differentiate the function u and y.

$$u = 3x^2 - x + 7$$

$$\frac{du}{dx} = 6x - 1$$

$$y = \frac{2}{(u)^3} = 2u^{-3}$$

$$\frac{dy}{du} = 2 \times (-3)u^{-3-1} = -6u^{-4}$$

$$\frac{dy}{dx} = \frac{dy}{du} \times \frac{du}{dx}$$

$$\frac{dy}{dx} = -6u^{-4} \times (6x - 1) = -6(6x - 1)u^{-4}$$

Recall $u = 3x^2 - x + 7$

$$\frac{dy}{dx} = -6(6x - 1)u^{-4} = -6(6x - 1)(3x^2 - x + 7)^{-4}$$

$$\frac{dy}{dx} = \frac{-6(6x - 1)}{(3x^2 - x + 7)^4}$$

Exercise 6a

---

1. Given that $y = (2x + 3)^4$, find $\frac{dy}{dx}$.

2. Given that $y = \dfrac{1}{(3x^2+1)^3}$, find $\dfrac{dy}{dx}$.

3. Find the derivative of $y = \sqrt{4x - 3}$ with respect to x.

4. Find the derivative of $y = \dfrac{1}{\sqrt{6x-5}}$ with respect to x.

5. Find the derivative of $y = \left(x - \dfrac{1}{x}\right)^3$ with respect to x.

Product Rule

In this case, we have

$$y = uv$$

Where u and v is a function of x

Apply increment

$$y + \Delta y = (u + \Delta u)(v + \Delta v)$$

Expand the equation

$$y + \Delta y = uv + u\Delta v + v\Delta u + \Delta u\Delta v$$

Subtract y from both side

$$y + \Delta y - y = uv + u\Delta v + v\Delta u + \Delta u\Delta v - y$$

Where $y = uv$

$$\Delta y = uv + u\Delta v + v\Delta u + \Delta u\Delta v - uv$$

$$\Delta y = u\Delta v + v\Delta u + \Delta u\Delta v$$

Find the gradient by dividing both sides by Δx

$$\frac{\Delta y}{\Delta x} = \frac{u\Delta v + v\Delta u + \Delta u\Delta v}{\Delta x}$$

$$\frac{\Delta y}{\Delta x} = \frac{u\Delta v}{\Delta x} + \frac{v\Delta u}{\Delta x} + \frac{\Delta u\Delta v}{\Delta x}$$

As $\Delta x \to 0, \Delta u \to 0, \Delta v \to 0$

Take $\lim\limits_{\Delta x \to 0} \frac{\Delta y}{\Delta x}$

$$\lim\limits_{\Delta x \to 0} \frac{f_{(x+\Delta x)} - f_{(x)}}{\Delta x} = \lim\limits_{\Delta x \to 0} \frac{u\Delta v}{\Delta x} + \textit{lim}\limits_{\Delta x \to 0} \frac{v\Delta u}{\Delta x} + \lim\limits_{\Delta x \to 0} \frac{\Delta u\Delta v}{\Delta x}$$

$$\textit{lim}\limits_{\Delta x \to 0} \frac{\Delta y}{\Delta x} = u \lim\limits_{\Delta x \to 0} \frac{\Delta v}{\Delta x} + v\,\textit{lim}\limits_{\Delta x \to 0} \frac{\Delta u}{\Delta x} + \Delta u \lim\limits_{\Delta x \to 0} \frac{\Delta v}{\Delta x}$$

$$\lim_{\Delta x \to 0} \frac{\Delta y}{\Delta x} = \frac{dy}{dx}$$

$$u \lim_{\Delta x \to 0} \frac{\Delta v}{\Delta x} = u \frac{dv}{dx}$$

$$v \lim_{\Delta x \to 0} \frac{\Delta u}{\Delta x} = v \frac{du}{dx}$$

$$\Delta u \lim_{\Delta x \to 0} \frac{\Delta v}{\Delta x} = \Delta u \frac{dv}{dx} = 0$$

Therefore

$$\frac{dy}{dx} = u \frac{dv}{dx} + v \frac{du}{dx}$$

Examples 1
Differentiate with respect to x; $y = (3 + 2x)(1 - x)$

<u>Solution</u>

$$y = (3 + 2x)(1 - x)$$

Let $\quad u = (3 + 2x)$

$\quad v = (1 - x)$

For if $y = ax^n$, $\frac{dy}{dx} = anx^{n-1}$

Differentiate the function u and v.

$$u = 3 + 2x, \quad \frac{du}{dx} = 2$$

$$v = 1 - x, \quad \frac{dv}{dx} = -1$$

$$\frac{dy}{dx} = u \frac{dv}{dx} + v \frac{du}{dx}$$

$$\frac{dy}{dx} = (3 + 2x)(-1) + (1 - x)(2)$$

$$\frac{dy}{dx} = (-3 - 2x) + (2 - 2x)$$

$$\frac{dy}{dx} = -3 - 2x + 2 - 2x = -1 - 4x$$

$$\frac{dy}{dx} = -(1 + 4x)$$

Examples 2
Differentiate with respect to x; $y = (2x + 1)(x + 1)$

Solution

$$y = (2x + 1)(x + 1)$$

Let $\quad u = 2x + 1$

$\quad\quad v = x + 1$

For if $y = ax^n$, $\frac{dy}{dx} = anx^{n-1}$

Differentiate the function u and v.

$$u = 2x + 1, \quad \frac{du}{dx} = 2$$

$$v = x + 1, \frac{dv}{dx} = 1$$

$$\frac{dy}{dx} = u\frac{dv}{dx} + v\frac{du}{dx}$$

$$\frac{dy}{dx} = (2x + 1)(1) + (x + 1)(2)$$

$$\frac{dy}{dx} = (2x + 1) + (2x + 1)$$

$$\frac{dy}{dx} = 2x + 1 + 2x + 1 = 4x + 2$$

$$\frac{dy}{dx} = 2(2x + 1)$$

Examples 3
Differentiate with respect to x; $y = (4 - 5x)(3 - 2x + 5x^2)$

<u>Solution</u>

$$y = (4 - 5x)(3 - 2x + 5x^2)$$

Let $\quad u = (4 - 5x)$

$\quad\quad v = (3 - 2x + 5x^2)$

$\quad\quad\quad$ For if $y = ax^n$, $\dfrac{dy}{dx} = anx^{n-1}$

Differentiate the function u and v.

$$u = 4 - 5x, \quad \frac{du}{dx} = -5$$

$$v = 3 - 2x + 5x^2, \frac{dv}{dx} = -2 + 10x$$

$$\frac{dy}{dx} = u\frac{dv}{dx} + v\frac{du}{dx}$$

$$\frac{dy}{dx} = (4 - 5x)(-2 + 10x) + (3 - 2x + 5x^2)(-5)$$

$$\frac{dy}{dx} = (-8 + 40x + 10x - 50x^2) + (-15 + 10x - 25x^2)$$

$$\frac{dy}{dx} = -8 - 15 + 40x + 10x + 10x - 50x^2 - 25x^2$$

$$\frac{dy}{dx} = -23 + 60x - 75x^2$$

Examples 4
Differentiate with respect to x; $y = (x^2 + 5)(3x - 1)^3$

<u>Solution</u>

$$y = (x^2 + 5)(3x - 1)^3$$

Let $\quad u = (x^2 + 5)$

$$v = (3x - 1)^3$$

For if $y = ax^n$, $\frac{dy}{dx} = anx^{n-1}$

Differentiate the function u and v.

$$u = x^2 + 5, \quad \frac{du}{dx} = 2x$$

$$v = (3x - 1)^3 \text{ using chain rule,}$$

$$\frac{dv}{dx} = 9(3x - 1)^2$$

$$\frac{dy}{dx} = u\frac{dv}{dx} + v\frac{du}{dx}$$

$$\frac{dy}{dx} = (x^2 + 5)9(3x - 1)^2 + (3x - 1)^3(2x)$$

$$\frac{dy}{dx} = (3x - 1)^2[9(x^2 + 5) + 2x(3x - 1)]$$

$$\frac{dy}{dx} = (3x - 1)^2[9x^2 + 45 + 6x^2 + 2x]$$

$$\frac{dy}{dx} = (3x - 1)^2[15x^2 + 2x + 45]$$

Examples 5

Differentiate with respect to x; $y = x\sqrt{x + 1}$

Solution

$$y = x\sqrt{x + 1}$$

Let $u = (x)$

$$v = \sqrt{x + 1} = (x + 1)^{\frac{1}{2}}$$

For if $y = ax^n$, $\frac{dy}{dx} = anx^{n-1}$

Differentiate the function u and v.

$$u = x, \quad \frac{du}{dx} = 1$$

$$v = (x + 1)^{\frac{1}{2}} \text{ using chain rule,}$$

$$\frac{dv}{dx} = \frac{1}{2}(x + 1)^{\frac{1}{2}-1} = \frac{1}{2}(x + 1)^{-\frac{1}{2}}$$

$$\frac{dy}{dx} = u\frac{dv}{dx} + v\frac{du}{dx}$$

$$\frac{dy}{dx} = (x)\frac{1}{2}(x + 1)^{-\frac{1}{2}} + 1(x + 1)^{\frac{1}{2}}$$

$$\frac{dy}{dx} = \frac{x}{\sqrt{x + 1}} + (\sqrt{x + 1}) = \frac{x(x + 1)}{\sqrt{x + 1}}$$

$$\frac{dy}{dx} = \frac{x(x + 1)}{\sqrt{x + 1}}$$

$$\frac{dy}{dx} = \frac{-6(6x - 1)}{(3x^2 - x + 7)^4}$$

Exercise 6b

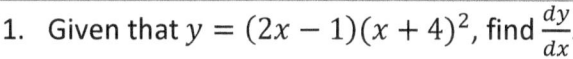

1. Given that $y = (2x - 1)(x + 4)^2$, find $\frac{dy}{dx}$.

2. Given that $y = x(x^2 - 1)^2$, find $\frac{dy}{dx}$.

3. Find the derivative of $y = (x^2 - 1)(x^3 + 1)$ with respect to x.

4. Find the derivative of $y = 3x^2(x^2 + 4)^2$ with respect to x.

5. Find the derivative of $y = \sqrt{x}(x + 3)^2$ with respect to x.

Quotient Rule

This is also known as the derivative of a quotient. Let say

$$y = \frac{u}{v}$$

Where u and v are function of x

$$y = \frac{u}{v}$$

Apply increment

$$y + \Delta y = \frac{u + \Delta v}{v + \Delta v}$$

Subtract y from both sides

$$y + \Delta y - y = \frac{u + \Delta v}{v + \Delta v} - y$$

$$\Delta y = \frac{u + \Delta v}{v + \Delta v} - y$$

Where

$$y = \frac{u}{v}$$

$$\Delta y = \frac{u + \Delta v}{v + \Delta v} - \frac{u}{v}$$

$$\Delta y = \frac{v(u + \Delta v) - u(v + \Delta v)}{v(v + \Delta v)}$$

$$\Delta y = \frac{v(u + \Delta v) - u(v + \Delta v)}{v(v + \Delta v)}$$

$$\Delta y = \frac{uv + v\Delta u - uv - u\Delta v}{v(v + \Delta v)}$$

$$\Delta y = \frac{v\Delta u - u\Delta v}{v(v + \Delta v)}$$

Find the gradient by dividing both sides by Δx

$$\frac{\Delta y}{\Delta x} = \frac{\frac{v\Delta u}{\Delta x} - \frac{u\Delta v}{\Delta x}}{v(v + \Delta v)}$$

As $\Delta x \to 0, \Delta u \to 0, \Delta v \to 0$

Take $\lim\limits_{\Delta x \to 0} \frac{\Delta y}{\Delta x} = \frac{dy}{dx}$

$$\lim\limits_{\Delta x \to 0} \frac{\Delta y}{\Delta x} = \lim\limits_{\Delta x \to 0} \frac{\frac{v\Delta u}{\Delta x} - \frac{u\Delta v}{\Delta x}}{v(v + \Delta v)}$$

$$= \frac{\lim\limits_{\Delta x \to 0} \frac{v\Delta u}{\Delta x} - \lim\limits_{\Delta x \to 0} \frac{u\Delta v}{\Delta x}}{v(v + \Delta v)}$$

Hence;

$$\frac{dy}{dx} = \frac{v\dfrac{du}{dx} - u\dfrac{dv}{dx}}{v^2}$$

Examples 1

Differentiate with respect to x; $y = \dfrac{(2+x^2)}{2-x^2}$

Solution

$$y = \frac{(2 + x^2)}{2 - x^2}$$

Let $\quad u = (2 + x^2)$

$$v = (2 - x^2)$$

For if $y = ax^n$, $\frac{dy}{dx} = anx^{n-1}$

Differentiate the function u and v.

$$u = 2 + x^2, \quad \frac{du}{dx} = 2x$$

$$v = 2 - x^2, \frac{dv}{dx} = -2x$$

$$\frac{dy}{dx} = \frac{v\dfrac{du}{dx} - u\dfrac{dv}{dx}}{v^2}$$

$$\frac{dy}{dx} = \frac{(2 - x^2)2x - (2 + x^2)(-2x)}{(2 - x^2)^2}$$

$$\frac{dy}{dx} = \frac{4x - 2x^3 - (-4x - 2x^3)}{(2 - x^2)^2}$$

$$\frac{dy}{dx} = \frac{4x - 2x^3 + 4x + 2x^3}{(2 - x^2)^2}$$

$$\frac{dy}{dx} = \frac{8}{(2 - x^2)^2}$$

Examples 2

Differentiate with respect to x; $y = \frac{(3 + 2x - x^2)}{\sqrt{1+x}}$

Solution

$$y = \frac{(3 + 2x - x^2)}{\sqrt{1 + x}}$$

Let $\quad u = (3 + 2x - x^2)$

$$v = \sqrt{1 + x}$$

For if $y = ax^n$, $\frac{dy}{dx} = anx^{n-1}$

Differentiate the function u and v.

$$u = 3 + 2x - x^2, \quad \frac{du}{dx} = 2 - 2x$$

$$v = \sqrt{1 + x} = (1 + x)^{\frac{1}{2}} \text{using chain rule,}$$

$$\frac{dv}{dx} = \frac{1}{2\sqrt{x + 1}}$$

$$\frac{dy}{dx} = \frac{v\frac{du}{dx} - u\frac{dv}{dx}}{v^2}$$

$$\frac{dy}{dx} = \frac{\sqrt{1 + x}(2 - 2x) - (3 + 2x - x^2)\frac{1}{2\sqrt{x + 1}}}{(\sqrt{1 + x})^2}$$

$$\frac{dy}{dx} = \frac{2(x + 1)(2 - 2x) - (3 + 2x - x^2)}{2(\sqrt{1 + x})(\sqrt{1 + x})^2}$$

$$\frac{dy}{dx} = \frac{2(2x - 2x^2 + 2 - 2x) - (3 + 2x - x^2)}{2(\sqrt{1 + x})x + 1}$$

$$\frac{dy}{dx} = \frac{4x^2 + 4 - 3 - 2x + x^2}{2(\sqrt{1 + x})x + 1}$$

$$\frac{dy}{dx} = \frac{5x^2 - 2x + 1}{2(\sqrt{1 + x})x + 1}$$

$$\frac{dy}{dx} = \frac{5x^2 - 2x + 1}{2x + 2(\sqrt{1 + x})}$$

Examples 3

Differentiate with respect to x; $y = \frac{(2+x)}{x^2+2x+7}$

Solution

$$y = \frac{(2+x)}{x^2 + 2x + 7}$$

Let $\quad u = (2 + x)$

$\quad v = (x^2 + 2x + 7)$

For if $y = ax^n$, $\frac{dy}{dx} = anx^{n-1}$

Differentiate the function u and v.

$$u = 2 + x, \quad \frac{du}{dx} = 1$$

$$v = x^2 + 2x + 7, \frac{dv}{dx} = 2x + 2$$

$$\frac{dy}{dx} = \frac{v\frac{du}{dx} - u\frac{dv}{dx}}{v^2}$$

$$\frac{dy}{dx} = \frac{(x^2 + 2x + 7)1 - (2+x)2x + 2}{(x^2 + 2x + 7)^2}$$

$$\frac{dy}{dx} = \frac{(x^2 + 2x + 7)1 - 4x + 4 + 2x^2 + 2x}{(x^2 + 2x + 7)^2}$$

$$\frac{dy}{dx} = \frac{x^2 + 2x + 7 - 6x + 4 + 2x^2}{(x^2 + 2x + 7)^2}$$

$$\frac{dy}{dx} = \frac{3x^2 - 4x + 1}{(x^2 + 2x + 7)^2}$$

$$\frac{dy}{dx} = \frac{3x^2 - 4x + 1}{(x^2 + 2x + 7)^2}$$

Examples 4

Differentiate with respect to x; $y = \frac{(\sqrt{x})}{x-2}$

Solution

$$y = \frac{(\sqrt{x})}{x-2}$$

Let $u = (\sqrt{x})$

$v = (x-2)$

For if $y = ax^n$, $\frac{dy}{dx} = anx^{n-1}$

Differentiate the function u and v.

$u = \sqrt{x}, \quad \frac{du}{dx} = \frac{1}{2\sqrt{x}}$

$v = x-2, \frac{dv}{dx} = 1$

$$\frac{dy}{dx} = \frac{v\frac{du}{dx} - u\frac{dv}{dx}}{v^2}$$

$$\frac{dy}{dx} = \frac{(x-2)\frac{1}{2\sqrt{x}} - (\sqrt{x})(1)}{(x-2)^2}$$

$$\frac{dy}{dx} = \frac{x-2-2x}{(x-2)^2}$$

$$\frac{dy}{dx} = \frac{-2-x}{(x-2)^2}$$

$$\frac{dy}{dx} = \frac{-(2+x)}{(x-2)^2}$$

Exercise 6c

1. Given that $y = \dfrac{x-1}{x+1}$, find $\dfrac{dy}{dx}$.

2. Given that $y = \frac{x^2+1}{x-1}$, find $\frac{dy}{dx}$.

3. Find the derivative of $y = \dfrac{x}{\sqrt{x-1}}$ with respect to x.

4. Find the derivative of $y = \dfrac{\left(2x^2-3\right)^3}{x}$ with respect to x.

5. Find the derivative of $y = \dfrac{\sqrt{x}}{x-2}$ with respect to x.

Implicit Differentiation

The process of differentiating implicit function is known as implicit differentiation. So far, all the functions we have differentiated have been expressed explicitly. y is said to be expressed explicitly in terms of x. The rules or technique of differentiation have used in this case is called explicit differentiation.

What of in case where a function may be given indirectly such that we have

$$xy + 2xy^2 + 3x = 0$$

In this case the relationship between x and y is implicit. The derivation of this function will be gotten by using implicit differentiation process.

Examples 1

Differentiate with respect to x; $x^3 + y^3 = 4$

Solution

To differentiate this function, each term will be separately differentiated

$$x^3 + y^3 = 4$$

The derivative of the function will be;

$$3x^{3-1} + 3y^{3-1}\frac{dy}{dx} = 0$$

$$3x^2 + 3y^2\frac{dy}{dx} = 0$$

$$3y^2\frac{dy}{dx} = 0 - 3x^2$$

$$3y^2 \frac{dy}{dx} = -3x^2$$

Divide both sides by $3y^2$

$$\frac{3y^2 \frac{dy}{dx}}{3y^2} = \frac{-3x^2}{3y^2}$$

$$\frac{dy}{dx} = \frac{-x^2}{y^2}$$

Examples 2

Differentiate with respect to x; $x^2y - 7x = 5$

Solution

To differentiate this function, each term will be separately differentiated

$$x^2y - 7x = 5$$

The derivative of each terms will be;

$$x^2y = 2xy + x^2 \frac{dy}{dx}$$

$$-7x = -7$$

$$5 = 0$$

Therefore;

$$2xy + x^2 \frac{dy}{dx} - 7 = 0$$

$$x^2 \frac{dy}{dx} = 0 - 2xy + 7$$

$$x^2 \frac{dy}{dx} = -2xy + 7$$

Divide both sides by x^2

$$\frac{x^2 \frac{dy}{dx}}{x^2} = \frac{-2xy + 7}{x^2}$$

$$\frac{dy}{dx} = \frac{-2xy + 7}{x^2}$$

Examples 3
Differentiate with respect to x; $x^2 + y^2 = 3xy$

Solution

To differentiate this function, each term will be separately differentiated

$$x^2 + y^2 = 3xy$$

The derivative of each terms will be;

$$x^2 = 2x$$

$$y^2 = 2y\frac{dy}{dx}$$

$$3xy = 3x\frac{dy}{dx} + 3y$$

Therefore;

$$2x + 2y\frac{dy}{dx} = 3x\frac{dy}{dx} + 3y$$

$$2y\frac{dy}{dx} - 3x\frac{dy}{dx} = 3y - 2x$$

$$(2y - 3x)\frac{dy}{dx} = -3y - 2x$$

Divide both sides by $(2y - 3x)$

$$\frac{(2y - 3x)\frac{dy}{dx}}{(2y - 3x)} = \frac{-3y - 2x}{(2y - 3x)}$$

$$\frac{dy}{dx} = \frac{-3y - 2x}{(2y - 3x)}$$

Examples 4

Differentiate with respect to x; $(x + y)^2 = 7$

Solution

To differentiate this function, each term will be separately differentiated

$$(x + y)^2 = 7$$

$$x^2 + 2xy + y^2 = 7$$

$$x^2 + y^2 = 3xy$$

The derivative of each terms will be;

$$x^2 = 2x$$

$$2xy = 2x\frac{dy}{dx} + 2y$$

$$y^2 = 2y\frac{dy}{dx}$$

$$7 = 0$$

Therefore;

$$2x + 2x\frac{dy}{dx} + 2y + 2y\frac{dy}{dx} = 0$$

$$2x\frac{dy}{dx} + 2y\frac{dy}{dx} = 0 - 2x - 2y$$

$$(2x + 2y)\frac{dy}{dx} = -2x - 2y$$

Divide both sides by $(2x + 2y)$

$$\frac{(2x + 2y)\frac{dy}{dx}}{(2x + 2y)} = \frac{-2x - 2y}{(2x + 2y)} = \frac{-2(x + y)}{2(x + y)} = -1$$

$$\frac{dy}{dx} = -1$$

Examples 5

Differentiate with respect to x; $x^2y + y^2x + 3x = 2$

Solution

To differentiate this function, each term will be separately differentiated

$$x^2y + y^2x + 3x = 2$$

The derivative of each terms will be;

$$x^2y = 2xy + x^2\frac{dy}{dx}$$

$$y^2x = 2xy\frac{dy}{dx} + y^2$$

$$3x = 3$$

$$2 = 0$$

Therefore;

$$2xy + x^2\frac{dy}{dx} + 2xy\frac{dy}{dx} + y^2 + 3 = 0$$

$$x^2\frac{dy}{dx} + 2xy\frac{dy}{dx} = 0 - 2xy - y^2 - 3$$

$$(x^2 + 2xy)\frac{dy}{dx} = -(2xy + y^2 + 3)$$

Divide both sides by $(x^2 + 2xy)$

$$\frac{(x^2 + 2xy)\frac{dy}{dx}}{(x^2 + 2xy)} = \frac{-(2xy + y^2 + 3)}{(x^2 + 2xy)}$$

$$\frac{dy}{dx} = \frac{-(2xy + y^2 + 3)}{(x^2 + 2xy)}$$

Exercise 6d

Find $\dfrac{dy}{dx}$ *in terms of x and y for these implicit functions*

1. $xy - x^2 = 3$, find $\dfrac{dy}{dx}$.

2. $x^2 + y^3 = 7$, find $\frac{dy}{dx}$.

3. $x^2 + 3xy = 7y$, find $\frac{dy}{dx}$.

4. $y(x^2 + 1) - x^2 = 3$, find $\frac{dy}{dx}$.

5. Find the derivative of $x^3 + y^3 = 9xy$ with respect to x.

Chapter 7
DERIVATIVE OF TRIGONOMETRIC FUNCTION

The differentiation of a trigonometric function is the mathematical process of finding the derivative of a trigonometric function, or its rate of change with respect to a variable.

In process of finding the derivative of the trigonometric functions, the following proofs are important.

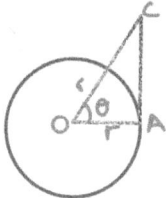

From the diagram above which shows a circle with centre O are radii

$OA = r$ and $OB = r$

Considering

Triangle OAB

sector OAB

Triangle OA

Let

Triangle OAB be A_1

Sector OAB be A_2

Triangle OAC be A_3

$$\text{Area } A_1 \text{ of Triangle OAB} = \frac{1}{2}ab\sin\theta$$

where $a \to |OA|, b \to |OB|$

$$A_1 = \frac{1}{2}|OA||OB|\sin\theta = \frac{1}{2}r^2\sin\theta$$

$$\text{Area } A_2 \text{ of Sector OAB} = \frac{\theta}{360} \times \pi r^2$$

$$= \frac{\theta}{2\pi}\pi r^2$$

$$A_2 = \frac{\theta}{2}r^2$$

$$\text{Area } A_3 \text{ of Triangle OAC} = \frac{1}{2} \times |OA| \times |AC|$$

$$A_3 = \frac{1}{2}\tan\theta$$

$$A_1 < A_2 < A_3$$

$$\frac{1}{2}r^2\sin\theta < \frac{\theta}{2}r^2 < \frac{1}{2}\tan\theta$$

Divide through by $\frac{1}{2}r^2\sin\theta$

$$1 < \frac{\theta}{\sin\theta} < \frac{1}{\cos\theta} \rightarrow 1 > \frac{\sin\theta}{\theta} > \frac{\cos\theta}{1}$$

Hence;

$\frac{\sin\theta}{\theta}$ must tend to 1 as θ tends to O from positive side.

Derivative of $sin\, x$

Let $y = \sin x$

Apply increment

$$y + \Delta y = \sin(x + \Delta x)$$

Subtract y from both sides

$$y + \Delta y - y = \sin(x + \Delta x) - y$$

$$\Delta y = \sin(x + \Delta x) - y$$

Where $y = \sin x$

$$\Delta y = \sin(x + \Delta x) - \sin x$$

From trigonometric identities (product formulae)

$$\sin A - \sin B = 2\cos\frac{A+B}{2}\sin\frac{A-B}{2}$$

$$\sin(x + \Delta x) - \sin x = 2\cos\frac{(x + \Delta x + x)}{2}\sin\frac{(x + \Delta x - x)}{2}$$

$$= 2\cos\frac{(2x + \Delta x)}{2}\sin\frac{(\Delta x)}{2}$$

$$\Delta y = \sin(x + \Delta x) - \sin x$$

$$\Delta y = 2\cos\frac{(2x + \Delta x)}{2}\sin\frac{(\Delta x)}{2}$$

$$\Delta y = 2\cos\left(x + \frac{\Delta x}{2}\right)\sin\frac{(\Delta x)}{2}$$

Find the gradient by dividing both sides byΔx

$$\frac{\Delta y}{\Delta x} = \frac{\cos\left(x + \frac{\Delta x}{2}\right)\sin\frac{(\Delta x)}{2}}{\frac{\Delta x}{2}}$$

$$As\ \Delta x \to 0 \quad \frac{\sin\frac{(\Delta x)}{2}}{\frac{\Delta x}{2}}$$

Take $\lim\limits_{\Delta x \to 0}\frac{\Delta y}{\Delta x}$

$$\frac{dy}{dx} = \cos x$$

Therefore;

$$if\ y = \sin x$$

$$\frac{dy}{dx} = \cos x$$

Derivative of $\cos x$

Let $y = \cos x$

it can be written that;

$$\cos x = \sin \frac{\pi}{2} - x$$

Using chain rule

Let
$$u = \frac{\pi}{2} - x$$

$$y = \sin u$$

Differentiate the function u and y.

$$u = \frac{\pi}{2} - x$$

$$\frac{du}{dx} = -1$$

$$y = \sin u$$

$$\frac{dy}{du} = \cos u$$

$$\frac{dy}{dx} = \frac{dy}{du} \times \frac{du}{dx}$$

$$\frac{dy}{dx} = \cos u \times -1 = -\cos u$$

Recall $u = \frac{\pi}{2} - x$

$$\frac{dy}{dx} = -\cos \frac{\pi}{2} - x$$

Since the answer must be in terms of x

$$-\cos \frac{\pi}{2} - x = -\sin x$$

Therefore

$$if\ y = \cos x$$

$$\frac{dy}{dx} = -\sin x$$

Derivative of tan x

Let $y = \tan x$

it can be written that;

$$\tan x = \frac{\sin x}{\cos x}$$

Using quotient rule

Let $u = \sin x$

$v = \cos x$

Differentiate the function u and v.

$$u = \sin x,\ \frac{du}{dx} = \cos x$$

$$v = \cos x,\ \frac{dv}{dx} = -\sin x$$

$$\frac{dy}{dx} = \frac{v\dfrac{du}{dx} - u\dfrac{dv}{dx}}{v^2}$$

$$\frac{dy}{dx} = \frac{(\cos x)\cos x - (\sin x)(-\sin x)}{(\cos x)^2}$$

$$\frac{dy}{dx} = \frac{\cos^2 x + \sin^2 x}{\cos^2 x}$$

From trigonometric identities

$$\cos^2 x + \sin^2 x = 1$$

Therefore;

$$\frac{dy}{dx} = \frac{1}{\cos^2 x} = \sec^2 x$$

$$\frac{dy}{dx} = \sec^2 x$$

If $y = \tan x$,

$$\frac{dy}{dx} = \sec^2 x$$

Derivative of sec x

Let $y = \sec x$

it can be written that;

$$\sec x = \frac{1}{\cos x}$$

Using chain rule

Let $\qquad u = \cos x$

$$y = \frac{1}{u}$$

Differentiate the function u and y.

$$u = \cos x$$

$$\frac{du}{dx} = -\sin x$$

$$y = \frac{1}{u} = u^{-1}$$

$$\frac{dy}{du} = -u^{-2}$$

$$\frac{dy}{dx} = \frac{dy}{du} \times \frac{du}{dx}$$

$$\frac{dy}{dx} = -u^{-2} \times -\sin x = -2u^{-2}(-\sin x)$$

Recall $u = \cos x$

$$\frac{dy}{dx} = \frac{1}{u^2}\sin x = \frac{\sin x}{u^2} = \frac{\sin x}{\cos^2 x}$$

$$\frac{\sin x}{\cos^2 x} = \frac{\sin x}{\cos x} \times \frac{1}{\cos x} = \tan x \sec x$$

$$\frac{dy}{dx} = \tan x \sec x$$

Derivative of csc x

Let $y = \csc x$

it can be written that;

$$\sec x = \frac{1}{\sin x}$$

Using chain rule

Let $\qquad u = \sin x$

$$y = \frac{1}{u}$$

Differentiate the function u and y.

$$u = \sin x$$

$$\frac{du}{dx} = \cos x$$

$$y = \frac{1}{u} = u^{-1}$$

$$\frac{dy}{du} = -u^{-2}$$

$$\frac{dy}{dx} = \frac{dy}{du} \times \frac{du}{dx}$$

$$\frac{dy}{dx} = -u^{-2} \times \cos x = -u^{-2}(\cos x)$$

Recall $u = \sin x$

$$\frac{dy}{dx} = \frac{-1}{u^2}\cos x = \frac{-\cos x}{\sin^2 x} = \frac{-\cos x}{\sin^2 x}$$

$$\frac{-\cos x}{\sin^2 x} = \frac{-\cos x}{\sin x} \times \frac{1}{\sin x} = -\cot x \csc x$$

$$\frac{dy}{dx} = -\cot x \csc x$$

Derivative of $\cot x$

Let $y = \cot x$

It can be written that;

$$\cot x = \frac{\cos x}{\sin x}$$

Using quotient rule

Let $\quad u = \cos x$

$\qquad v = \sin x$

Differentiate the function u and v.

$$u = \cos x, \quad \frac{du}{dx} = -\sin x$$

$$v = \sin x, \quad \frac{dv}{dx} = \cos x,$$

$$\frac{dy}{dx} = \frac{v\frac{du}{dx} - u\frac{dv}{dx}}{v^2}$$

$$\frac{dy}{dx} = \frac{(\sin x)(-\sin x) - (\cos x)(\cos x)}{(\cos x)^2}$$

$$\frac{dy}{dx} = \frac{-\sin^2 x - \cos^2 x}{\sin^2 x} = \frac{-(\sin^2 x + \cos^2 x)}{\sin^2 x}$$

From trigonometric identities

$$\sin^2 x + \cos^2 x = 1$$

Therefore;

$$\frac{dy}{dx} = \frac{-1}{\sin^2 x} = -\csc^2 x$$

$$\frac{dy}{dx} = -\csc^2 x$$

If $y = \cot x$,

$$\frac{dy}{dx} = -\csc^2 x$$

The table below summarizes the derivatives of trigonometric functions

TRIGONOMETRIC FUNCTIONS	DERIVATIVES
$\sin x$	$\cos x$
$\cos x$	$-\sin x$
$\tan x$	$\sec^2 x$
$\csc x$	$-\csc x \cot x$
$\sec x$	$-\sec x \cot x$
$\cot x$	$-\csc^2 x$

Using the table above, the derivative of all other trigonometric function can be resolved.

Examples 1
Find the derivative of $\cos 2x$

<u>Solution</u>

Let $y = \cos 2x$

Using chain rule

Let
$$u = 2x$$
$$y = \cos x$$
Differentiate the function u and y.

$$u = 2x$$

$$\frac{du}{dx} = 2$$

$$y = \cos u$$

$$\frac{dy}{du} = -\sin u$$

$$\frac{dy}{dx} = \frac{dy}{du} \times \frac{du}{dx}$$

$$\frac{dy}{dx} = -\sin u \times 2 = -2\sin u$$

Recall u = 2x

$$\frac{dy}{dx} = -2\sin 2x$$

Therefore

$$if\ y = \cos 2x$$

$$\frac{dy}{dx} = -2\sin 2x$$

Examples 2

Find the derivative of $sin\frac{1}{3}x$

<u>Solution</u>

Let $y = \sin\frac{1}{3}x$

Using chain rule

Let $$u = \frac{1}{3}x$$

$$y = \sin u$$

Differentiate the function u and y.

$$u = \frac{1}{3}x$$

$$\frac{du}{dx} = \frac{1}{3}$$

$$y = \sin u$$

$$\frac{dy}{du} = \cos u$$

$$\frac{dy}{dx} = \frac{dy}{du} \times \frac{du}{dx}$$

$$\frac{dy}{dx} = \cos u \times \frac{1}{3} = \frac{1}{3}\cos u$$

Recall $u = \frac{1}{3}x$

$$\frac{dy}{dx} = \frac{1}{3}\cos\frac{1}{3}x$$

Therefore

$$if \ y = \sin\frac{1}{3}x$$

$$\frac{dy}{dx} = \frac{1}{3}\cos\frac{1}{3}x$$

Examples 3
Find the derivative of $\cos^2 x$

Solution

Let $y = \cos^2 x$

Using chain rule

Let $\qquad\qquad u = \cos x$

$$y = u^2$$

Differentiate the function u and y.

$$u = \cos x$$

$$\frac{du}{dx} = -\sin x$$

$$y = u^2$$

$$\frac{dy}{du} = 2u$$

$$\frac{dy}{dx} = \frac{dy}{du} \times \frac{du}{dx}$$

$$\frac{dy}{dx} = 2u \times (-\sin x) = -2u \sin x$$

Recall $u = \cos x$

$$\frac{dy}{dx} = -2 \cos x \sin x$$

Therefore

$$if \ \ y = \cos^2 x$$

$$\frac{dy}{dx} = -2 \cos x \sin x$$

Examples 4

Find the derivative of $\csc \frac{3}{4} x^2$

Solution

Let $y = \csc \frac{3}{4} x^2$

Using chain rule

Let

$$u = \frac{3}{4} x^2$$

$$y = \csc u$$

Differentiate the function u and y.

$$u = \frac{3}{4}x^2$$

$$\frac{du}{dx} = \frac{3}{2}x$$

$$y = \csc u$$

$$\frac{dy}{du} = -\csc u \cot u$$

$$\frac{dy}{dx} = \frac{dy}{du} \times \frac{du}{dx}$$

$$\frac{dy}{dx} = -\csc x \cot x \times \left(\frac{3}{2}x\right) = \frac{-3}{2}x \csc u \cot u$$

Recall $u = \frac{3}{4}x^2$

$$\frac{dy}{dx} = \frac{-3}{2}x \csc\frac{3}{4}x^2 \cot\frac{3}{4}x^2$$

Therefore

$$if \ y = \csc\frac{3}{4}x^2$$

$$\frac{dy}{dx} = \frac{dy}{dx} = \frac{-3}{2}x \csc\frac{3}{4}x^2 \cot\frac{3}{4}x^2$$

Examples 5
Find the derivative of $\tan^2 x^2$

<u>Solution</u>

Let $y = \tan^2 x^2$

Using chain rule

Let

$$v = x^2$$

$$u = \tan v$$

$$y = u^2$$

Differentiate the function u, v and y.

$$v = x^2$$

$$\frac{dv}{dx} = 2x$$

$$u = \tan v$$

$$\frac{du}{dv} = \sec^2 v$$

$$y = u^2$$

$$\frac{dy}{du} = 2u$$

$$\frac{dy}{dx} = \frac{dy}{du} \times \frac{du}{dv} \times \frac{dv}{dx}$$

$$\frac{dy}{dx} = 2u \times (\sec^2 v) \times 2x = 4xu \sec^2 v$$

Recall $u = \tan v$

$$v = x^2$$

$$\frac{dy}{dx} = 4xu \sec^2 v = 4x \tan x^2 \sec^2 x^2$$

Therefore

$$if \ y = \tan^2 x^2$$

$$\frac{dy}{dx} = 4x \tan x^2 \sec^2 x^2$$

Examples 6

Find the derivative of $\tan \sqrt{x}$

Solution

Let $y = \tan \sqrt{x}$

Apply chain rule

Let $$u = \sqrt{x} = x^{\frac{1}{2}}$$

$$y = \tan u$$

Differentiate the function u and y.

$$u = x^{\frac{1}{2}}$$

$$\frac{du}{dx} = \frac{1}{2}x^{-\frac{1}{2}} = \frac{1}{2\sqrt{x}}$$

$$y = \tan u$$

$$\frac{dy}{du} = \sec^2 u$$

$$\frac{dy}{dx} = \frac{dy}{du} \times \frac{du}{dx}$$

$$\frac{dy}{dx} = \sec^2 u \times \frac{1}{2\sqrt{x}} = \frac{\sec^2 u}{2\sqrt{x}}$$

Recall $$u = \sqrt{x}$$

$$\frac{dy}{dx} = \frac{\sec^2 \sqrt{x}}{2\sqrt{x}}$$

Therefore

$$if \ \ y = \tan\sqrt{x}$$

$$\frac{dy}{dx} = \frac{\sec^2 \sqrt{x}}{2\sqrt{x}}$$

Examples 7
Find the derivative of $x \sin x$

Solution

Let $y = x \sin x$

Apply product rule

Let $\qquad u = x$

$$v = \sin x$$

Differentiate the function u and v.

$$u = x^{\frac{1}{2}}, \qquad\qquad \frac{du}{dx} = 1$$

$$v = \sin x, \qquad\qquad \frac{dv}{dx} = \cos x$$

$$\frac{dy}{dx} = u\frac{dv}{dx} + v\frac{du}{dx}$$

$$\frac{dy}{dx} = x(\cos x) + (\sin x)1$$

$$\frac{dy}{dx} = x\cos x + \sin x$$

Therefore

$$if\, y = x\sin x$$

$$\boldsymbol{\frac{dy}{dx} = x\cos x + \sin x}$$

Examples 8
Find the derivative of $\cos(2x - 1)$

Solution

Let $y = \cos(2x - 1)$

Apply chain rule

Let $\qquad u = 2x - 1$

$$y = \cos u$$

Differentiate the function u and y.

$$u = 2x - 1, \qquad\qquad \frac{du}{dx} = 2$$

$$y = \cos u, \qquad \frac{dy}{du} = -\sin u$$

$$\frac{dy}{dx} = \frac{dy}{du} \times \frac{du}{dx}$$

$$\frac{dy}{dx} = -\sin u \times 2 = -2 \sin u$$

Recall
$$u = 2x - 1$$

$$\frac{dy}{dx} = -2 \sin(2x - 1)$$

Therefore

$$if \ \ y = \cos(2x - 1)$$

$$\frac{dy}{dx} = -2 \sin(2x - 1)$$

Examples 9
Find the derivative of $\tan 4x^2$

Solution

Let $y = \tan 4x^2$

Apply chain rule

Let
$$u = 4x^2$$

$$y = \tan u$$

Differentiate the function u and y.

$$u = 4x^2, \qquad \frac{du}{dx} = 8x$$

$$y = \tan u, \qquad \frac{dy}{du} = \sec^2 u$$

$$\frac{dy}{dx} = \frac{dy}{du} \times \frac{du}{dx}$$

$$\frac{dy}{dx} = \sec^2 u \times 8x = 8x \sec^2 u$$

Recall
$$u = 4x^2$$

$$\frac{dy}{dx} = 8x \sec^2 4x^2$$

Therefore

$$if \ \ y = \tan 4x^2$$

$$\frac{dy}{dx} = 8x \sec^2 4x^2$$

Examples 10

Find the derivative of $\frac{1+\cos 2x}{\sin 2x}$

Solution

Let $y = \frac{1+\cos 2x}{\sin 2x}$

Apply quotient rule

Let
$$u = 1 + \cos 2x$$

$$v = \sin 2x$$

Differentiate the function $u \ and \ v$.

$$u = 1 + \cos 2x, \qquad \frac{du}{dx} = -2\sin 2x$$

$$v = \sin 2x \qquad \frac{dv}{dx} = 2\cos 2x$$

$$\frac{dy}{dx} = \frac{v\dfrac{du}{dx} - u\dfrac{dv}{dx}}{v^2}$$

$$\frac{dy}{dx} = \frac{\sin 2x \, (-2\sin 2x) - (1 + \cos 2x)(2\cos 2x)}{\sin^2 2x}$$

$$\frac{dy}{dx} = \frac{(-2\sin^2 2x) - (2\cos 2x + 2\cos^2 2x)}{\sin^2 2x}$$

$$\frac{dy}{dx} = \frac{-2\sin^2 2x - 2\cos 2x - 2\cos^2 2x}{\sin^2 2x}$$

$$\frac{dy}{dx} = \frac{-2(\sin^2 2x + \cos 2x + \cos^2 2x)}{\sin^2 2x}$$

If $\sin^2 \theta + \cos^2 \theta = 1$

$$\frac{dy}{dx} = \frac{-2(\sin^2 2x + \cos^2 2x + \cos 2x)}{\sin^2 2x}$$

$$\frac{dy}{dx} = \frac{-2(1 + \cos 2x)}{\sin^2 2x}$$

Therefore

$$if\, y = \frac{1 + \cos 2x}{\sin 2x}$$

$$\frac{dy}{dx} = \frac{-2(1 + \cos 2x)}{\sin^2 2x}$$

Exercise 7a

Find the derivative of the following trigonometric functions.

1. $y = \sin \sqrt{x}$

2. $y = x \cos x.$

3. $y = \cot(2 + x)$

4. $y = \tan 4\,x^2$

5. $y = \frac{1}{x}\sec x$

THE DERIVATIVE OF INVERSE FUNCTION

Let $y = f_{(x)}$

Then $x = f_{(y)}^{-1}$

$$\frac{dx}{dy} = \frac{1}{dy/dx}$$

$$\frac{dy}{dx} = \frac{1}{dx/dy}$$

Examples 1

If $y = \sqrt[3]{x}$, find $\frac{dy}{dx}$.

Solution

$$y^3 = x$$
$$x = y^3$$
$$\frac{dy}{dx} = 3y^2$$

$$\frac{dy}{dx} = 3y^2$$

THE DERIVATIVE OF LOGARITHM FUNCTION

The logarithm to the base of "e" of a number is called the natural

logarithm of that number. $\log_e x = \ln x$

Take note of the following derivatives to find the derivatives of logarithm

function.

The derivative of $\log_e x$

If $y = \log_e x$

$$\frac{dy}{dx} = \frac{1}{x}$$

The derivative of $\log_a x$

If $y = \log_a e$

$$\frac{dy}{dx} = \frac{1}{x}\log_a e$$

Examples 1

Find the derivative of the following.

$$\log_a(3x + 2)$$

Solution

$$y = \log_a(3x + 2)$$

Let $u = 3x + 2$

$$y = \log_a u$$

Differentiate y *and* u

$$u = 3x + 2$$

$$\frac{du}{dx} = 3$$

$$y = \log_a u$$

$$\frac{dy}{du} = \frac{1}{u}\log_a e$$

$$\frac{dy}{dx} = \frac{dy}{du} \times \frac{du}{dx}$$

$$\frac{dy}{dx} = \frac{1}{u} \log_a e \times 3$$

Where $u = 3x + 2$

$$\frac{dy}{dx} = \frac{3}{3x + 2} \log_a e$$

Examples 2

Find the derivative of the following $\log_e(2x - 1)^2$

Solution

$$y = \log_e(2x - 1)^2 = \ln(2x - 1)^2$$

Let $u = (2x - 1)^2$

$$y = \log_e u$$

Differentiate y and u

$$u = (2x - 1)^2$$

$$\frac{du}{dx} = 4(2x - 1)$$

$$y = \log_e u$$

$$\frac{dy}{du} = \frac{1}{u}$$

$$\frac{dy}{dx} = \frac{dy}{du} \times \frac{du}{dx}$$

$$\frac{dy}{dx} = \frac{1}{u} \times 4(2x - 1)$$

Where $u = (2x - 1)^2$

$$\frac{dy}{dx} = \frac{4(2x - 1)}{(2x - 1)^2} = \frac{4}{2x - 1}$$

$$\frac{dy}{dx} = \frac{4}{2x-1}$$

Examples 3
Find the derivative of the following.

$$\log_a(5x+3)$$

Solution

$$y = \log_a(5x+3)$$

Let $u = 5x + 3$

$$y = \log_a u$$

Differentiate y and u

$$u = 5x + 3$$

$$\frac{du}{dx} = 5$$

$$y = \log_a u$$

$$\frac{dy}{du} = \frac{1}{u}\log_a e$$

$$\frac{dy}{dx} = \frac{dy}{du} \times \frac{du}{dx}$$

$$\frac{dy}{dx} = \frac{1}{u}\log_a e \times 5$$

Where $u = 5x + 3$

$$\frac{dy}{dx} = \frac{5}{5x+3}\log_a e$$

Examples 4

Find the derivative of the following $\log_e \sqrt{2+x}$

Solution

$$y = \log_e \sqrt{2+x} = \ln\sqrt{2+x}$$

Let $u = \sqrt{2+x}$

$$y = \log_e u$$

Differentiate $y \ and \ u$

$$u = (2+x)^{\frac{1}{2}}, \qquad\qquad \frac{du}{dx} = \frac{1}{2}(2+x)^{-\frac{1}{2}} = \frac{1}{2\sqrt{2+x}}$$

$$y = \log_e u, \qquad\qquad \frac{dy}{du} = \frac{1}{u}$$

$$\frac{dy}{dx} = \frac{dy}{du} \times \frac{du}{dx}$$

$$\frac{dy}{dx} = \frac{1}{u} \times \frac{1}{2\sqrt{2+x}}$$

Where $u = \sqrt{2+x}$

$$\frac{dy}{dx} = \frac{1}{\sqrt{2+x}} \times \frac{1}{2\sqrt{2+x}}$$

$$\frac{dy}{dx} = \frac{1}{2(2+x)}$$

Examples 5

Find the derivative of the following $\log_e \frac{1}{x}$

Solution

$$y = \log_e \frac{1}{x} = \ln \frac{1}{x}$$

Let $u = \frac{1}{x}$

$$y = \log_e u$$

Differentiate y and u

$$u = \frac{1}{x} = x^{-1}, \qquad \frac{du}{dx} = -x^{-2} = \frac{-1}{x^2}$$

$$y = \log_e u, \qquad \frac{dy}{du} = \frac{1}{u}$$

$$\frac{dy}{dx} = \frac{dy}{du} \times \frac{du}{dx}$$

$$\frac{dy}{dx} = \frac{1}{u} \times \frac{-1}{x^2}$$

Where $u = \frac{1}{x}$

$$\frac{dy}{dx} = \frac{1}{x} \times \frac{-1}{x^2} = -(x^{-1} \times x^{-2}) = -x^{-3}$$

$$\frac{dy}{dx} = \frac{-1}{x^3}$$

Exercise 7b

Find the derivative of the following trigonometric functions.
  1. $y = \ln(1 + x)^4$

2. $y = \dfrac{\log_e x}{1+\sin x}$.

3. $y = x^2 \log_e x$

4. $y = \log_e(1 + 2x)^2$

5. $y = \log_a 3x$

THE DERIVATIVE OF EXPONENTIAL FUNCTION

Take note of the following derivatives to find the derivatives of exponential function.

The derivative of a^x

If $y = a^x$

$$\frac{dy}{dx} = \log_e a^x$$

The derivative of e^x

If $y = e^x$

$$\frac{dy}{dx} = e^x$$

Examples 1

Find the derivative of e^{2x}

$$y = e^{2x}$$

Solution

Apply chain rule

Let $u = 2x$

$$y = e^u$$

Differentiate y *and* u

$$u = 2x$$

$$\frac{du}{dx} = 2$$

$$y = e^u$$

$$\frac{dy}{du} = e^u$$

$$\frac{dy}{dx} = \frac{dy}{du} \times \frac{du}{dx}$$

$$\frac{dy}{dx} = e^u \times 2 = 2e^u$$

Where $u = 2x$

$$\frac{dy}{dx} = 2e^u = 2e^{2x}$$

$$\frac{dy}{dx} = 2e^{2x}$$

Examples 2

Find the derivative of e^{-4x-5}: $y = e^{-4x-5}$

<u>Solution</u>

Apply chain rule

Let $u = -4x - 5$

$$y = e^u$$

Differentiate y *and* u

$$u = -4x - 5$$

$$\frac{du}{dx} = -4$$

$$y = e^u$$

$$\frac{dy}{du} = e^u$$

$$\frac{dy}{dx} = \frac{dy}{du} \times \frac{du}{dx}$$

$$\frac{dy}{dx} = e^u \times -4 = -4e^u$$

Where $u = -4x - 5$

$$\frac{dy}{dx} = -4e^u = -4e^{-4x-5}$$

$$\frac{dy}{dx} = -4e^{-4x-5}$$

Examples 3
Find the derivative of $e^{\cos x}$: $\qquad y = e^{\cos x}$

Solution

Apply chain rule

Let $\qquad\qquad\qquad u = \cos x$

$$y = e^u$$

Differentiate y and u

$$u = \cos x \qquad\qquad \frac{du}{dx} = -\sin x$$

$$y = e^u \qquad\qquad \frac{dy}{du} = e^u$$

$$\frac{dy}{dx} = \frac{dy}{du} \times \frac{du}{dx}$$

$$\frac{dy}{dx} = e^u \times -\sin x = e^u(-\sin x)$$

Where $u = \cos x$

$$\frac{dy}{dx} = e^u(-\sin x) = -\sin x \, e^{\cos x}$$

$$\frac{dy}{dx} = -\sin x \, e^{\cos x}$$

Examples 4

Find the derivative of $e^{\ln x}$: $y = e^{\ln x}$

<u>Solution</u>

Apply chain rule

Let

$$u = \ln x$$

$$y = e^u$$

Differentiate y and u

$$u = \ln x \qquad\qquad \frac{du}{dx} = \frac{1}{x}$$

$$y = e^u \qquad\qquad \frac{dy}{du} = e^u$$

$$\frac{dy}{dx} = \frac{dy}{du} \times \frac{du}{dx}$$

$$\frac{dy}{dx} = e^u \times \frac{1}{x}$$

Where $u = \ln x$

$$\frac{dy}{dx} = e^u \left(\frac{1}{x}\right) = \frac{e^{\ln x}}{x}$$

$$\frac{dy}{dx} = \frac{e^{\ln x}}{x}$$

Examples 5

Find the derivative of $2e^x \sqrt{x}$: $y = 2e^x \sqrt{x}$

Solution

Let $y = 2e^x\sqrt{x}$

Apply product rule

Let $\qquad\qquad u = 2e^x$

$$v = \sqrt{x}$$

Differentiate the function u and v.

$$u = 2e^x \qquad\qquad \frac{du}{dx} = 2e^x$$

$$v = \sqrt{x}, \qquad\qquad \frac{dv}{dx} = \frac{1}{2\sqrt{x}}$$

$$\frac{dy}{dx} = u\frac{dv}{dx} + v\frac{du}{dx}$$

$$\frac{dy}{dx} = 2e^x\left(\frac{1}{2\sqrt{x}}\right) + (\sqrt{x})2e^x$$

$$\frac{dy}{dx} = 2e^x\left(\frac{1}{2\sqrt{x}} + \sqrt{x}\right)$$

Therefore

$$if\, y = 2e^x\sqrt{x}$$

$$\frac{dy}{dx} = 2e^x\left(\frac{1 + 2x}{2\sqrt{x}}\right)$$

Exercise 7c

Find the derivative of the following trigonometric functions.

  1. $y = e^{-2x-3}$

2. $y = e^{\sqrt{x}}$

3. $y = x^2 e^{2x}$

4. $y = e^{\sin x - \cos x}$

5. $y = e^{(x+3)^2}$

Chapter 8

HIGHER DERIVATIVE

Given that

$$y = f_{(x)}$$

$$\frac{dy}{dx} = f'_{(x)}$$

You can further differentiate $\frac{dy}{dx}$, and the result becomes $\frac{d}{dx}\left[\frac{dy}{dx}\right] \cdot \frac{d}{dx}\left[\frac{dy}{dx}\right]$ is

called the second derivative of y with respect to x.

Second derivative is denoted by $\frac{d^2y}{dx^2}$. The differentiation can continue,

thereby having third, fourth, fifth to n^{th} derivative.

The third, fourth and fifth derivatives are represented by

$\frac{d^3y}{dx^3}, \frac{d^4y}{dx^4}, \frac{d^5y}{dx^5} \cdots \frac{d^ny}{dx^n}$.

These higher derivatives can take the following representations.

$$\frac{dy}{dx} = f'_{(x)}$$

$$\frac{d^2y}{dx^2} = f''_{(x)}$$

$$\frac{d^2y}{dx^2} = f'''_{(x)}$$

$$\frac{d^4y}{dx^4} = f^{iv}_{(x)}$$

$$\frac{d^5y}{dx^5} = f^{v}_{(x)}$$

.

.

.

$$\frac{d^n y}{dx^n} = f^n_{(x)}$$

Examples 1
Find the first and second derivative derivative of the following.
$$y = 4x^5$$

Solution

$$y = 4x^5$$

if $y = ax^n, \frac{dy}{dx} = anx^{n-1}$

Therefore;

First derivative $\frac{dy}{dx}$

$$\frac{dy}{dx} = 5 \times 4x^{5-1} = 20x^4$$

$$\frac{dy}{dx} = 20x^4$$

Second derivative $\frac{d^2 y}{dx^2}$

$$\frac{d^2 y}{dx^2} = \frac{d}{dx}\left[\frac{dy}{dx}\right]$$

$$\frac{d^2 y}{dx^2} = \frac{d}{dx}[20x^4]$$

$$\frac{d^2 y}{dx^2} = 4 \times 20x^{4-1} = 80x^3$$

$$\frac{d^2 y}{dx^2} = 80x^3$$

Examples 2

Find the first and second derivative derivative of the following.

$$y = 3x^5 - 2x^4 + 5x^3 - x^2 + 3$$

Solution

$$y = 3x^5 - 2x^4 + 5x^3 - x^2 + 3$$

if $y = ax^n, \dfrac{dy}{dx} = anx^{n-1}$

Therefore;

First derivative $\dfrac{dy}{dx}$

$$\frac{dy}{dx} = 5 \times 3x^{5-1} - 4 \times 2x^{4-1} + 3 \times 5x^{3-1} - 2x^{2-1}$$

$$\frac{dy}{dx} = 15x^4 - 8x^3 + 15x^2 - 2x$$

Second derivative $\dfrac{d^2y}{dx^2}$

$$\frac{d^2y}{dx^2} = \frac{d}{dx}\left[\frac{dy}{dx}\right]$$

$$\frac{d^2y}{dx^2} = \frac{d}{dx}[15x^4 - 8x^3 + 15x^2 - 2x]$$

$$\frac{d^2y}{dx^2} = 4 \times 15x^{4-1} - 3 \times 8x^{3-1} + 2 \times 15x^{2-1} - 1 \times 2x^{1-1}$$

$$\frac{d^2y}{dx^2} = 60x^3 - 24x^2 + 30x^1 - 2x^0$$

$$\frac{d^2y}{dx^2} = 60x^3 - 24x^2 + 30x - 2$$

Examples 3

Find the first and second derivative derivative of the following.

$$y = \ln x$$

Solution

$$y = \ln x$$

First derivative $\left(\frac{dy}{dx}\right)$

Let $u = x$

$$y = \log_e u$$

Differentiate y and u

$$u = x, \qquad \frac{du}{dx} = 1$$

$$y = \log_e u, \qquad \frac{dy}{du} = \frac{1}{u}\log_e e = \frac{1}{u}$$

$$\frac{dy}{dx} = \frac{dy}{du} \times \frac{du}{dx}$$

$$\frac{dy}{dx} = \frac{1}{u} \times 1$$

Where $u = x$

$$\frac{dy}{dx} = \frac{1}{x} \times 1 = \frac{1}{x}$$

$$\boldsymbol{\frac{dy}{dx}} = \frac{1}{x} = x^{-1}$$

Second derivative $\frac{d^2y}{dx^2}$

if $y = ax^n, \frac{dy}{dx} = anx^{n-1}$

$$\frac{d^2y}{dx^2} = \frac{d}{dx}\left[\frac{dy}{dx}\right]$$

$$\frac{d^2y}{dx^2} = \frac{d}{dx}[x^{-1}]$$

$$\frac{d^2y}{dx^2} = -1 \times x^{-1-1} = -x^{-2}$$

$$\frac{d^2y}{dx^2} = -x^{-2}$$

Examples 4

Find the first and second derivative derivative of the following.

$$y = \sin x$$

Solution

$$y = \sin x$$

First derivative $\left(\frac{dy}{dx}\right)$

$$\frac{dy}{dx} = \cos x$$

$$\frac{dy}{dx} = \cos x$$

Second derivative $\quad \frac{d^2y}{dx^2}$

$$\frac{d^2y}{dx^2} = \frac{d}{dx}\left[\frac{dy}{dx}\right]$$

$$\frac{d^2y}{dx^2} = \frac{d}{dx}[\cos x]$$

$$\frac{d^2y}{dx^2} = -\sin x = -\sin x$$

$$\frac{d^2y}{dx^2} = -\sin x$$

Examples 5

Find the first and second derivative derivative of the following.

$$y = e^{x^3}$$

Solution

$$y = \ln x$$

First derivative $\left(\frac{dy}{dx}\right)$

Let $u = x^3$

$$y = e^u$$

Differentiate y *and* u

$$u = x^3, \qquad\qquad \frac{du}{dx} = 3x^2$$

$$y = e^u, \qquad\qquad \frac{dy}{du} = e^u$$

$$\frac{dy}{dx} = \frac{dy}{du} \times \frac{du}{dx}$$

$$\frac{dy}{dx} = e^u \times 3x^2$$

Where $u = x^3$

$$\frac{dy}{dx} = 3x^2 \times e^{x^3} = 3x^2 e^{x^3}$$

$$\frac{dy}{dx} = 3x^2 e^{x^3}$$

Second derivative $\frac{d^2y}{dx^2}$

$$\frac{d^2y}{dx^2} = \frac{d}{dx}\left[\frac{dy}{dx}\right]$$

$$\frac{d^2y}{dx^2} = \frac{d}{dx}\left[3x^2 e^{x^3}\right]$$

Apply product rule

Let $\qquad\qquad u = 3x^2$

$$v = e^{x^3}$$

Differentiate the function u and v.

$$u = 3x^2, \qquad\qquad \frac{du}{dx} = 6x$$

$$v = e^{x^3}, \qquad\qquad \frac{dv}{dx} = 3x^2 e^{x^3}$$

$$\frac{d^2y}{dx^2} = u\frac{dv}{dx} + v\frac{du}{dx}$$

$$\frac{d^2y}{dx^2} = 3x^2\left(3x^2 e^{x^3}\right) + \left(e^{x^3}\right)6x$$

$$\frac{d^2y}{dx^2} = 9x^4 e^{x^3} + 6x e^{x^3}$$

$$\frac{d^2y}{dx^2} = e^{x^3}(9x^4 + 6x)$$

Exercise 8

1. **If** $y = x^3 + 3x^2$, **Find** $\frac{dy}{dx}, \frac{d^2y}{dx^2}, \frac{d^3y}{dx^3}$.

2. Find the second derivative of $\tan x^3$

3. Find the second derivative of e^{x^2}

4. If $y = \sin 2x$, find $\dfrac{d^2y}{dx^2}$

5. Find $\frac{d^2y}{dx^2}$, if $y = \frac{1}{x+1}$

Chapter 9

CALCULUS II: APPLICATION OF DIFFERENTIATION

Having understood how we can differentiate various functions. This chapter will concentrate more on how derivative of function can be applied in different aspect of mathematics.

Differentiation can be applied in the following aspect of mathematics;

1. Tangent and Normal
2. Increasing and Decreasing function.
3. Approximation
4. Rate of change.
5. Rectilinear motion
6. Maximum and minimum values.

Tangent And Normal

The tangent is a straight line which touches the curve at a given point. The Normal is a straight line which is perpendicular to the tangent.

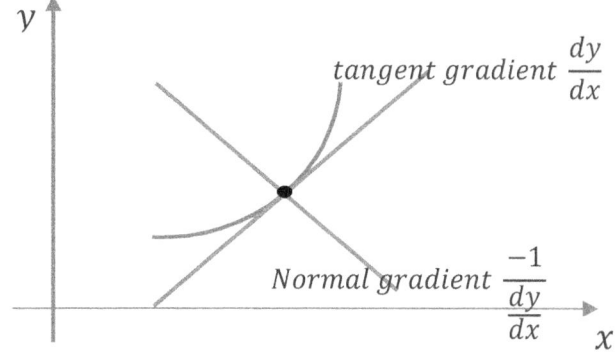

From the diagram above, the straight line perpendicular to the tangent at the point of contact of the tangent to the curve is called Normal to the curve.

If the gradient of the tangent at the point of contact to the point of contact to the curve is $m = \frac{dy}{dx}$, and the gradient of the normal is m_1, hence the gradient of the line perpendicular to the tangent, **the Normal to the curve is**

$$m_1 = \frac{-1}{m} = \frac{-1}{\dfrac{dy}{dx}}$$

Then, at a given co-ordinate point (x_1, y_1) the equation of the normal will be

$$y - y_1 = \frac{-1(x - x_1)}{m_1}$$

Examples 1

Find the equation of the tangent and normal to the curve $y = (x + 3)(x + 2)$ *at the point where* $x = 2$.

Solution

$$y = (x + 3)(x + 2)$$

How to find the equation of tangent and the equation of Normal to the curve using derivative of the function.

$$y = x^2 + 2x + 3x + 6$$
$$y = x^2 + 5x + 6$$

At point $x = 2$

$$y = 2^2 + 5(2) + 6$$

$$y = 4 + 10 + 6 = 20$$

$$y = 20$$

Therefore the co-ordinate points are $(2, 20)$

The derivative of function y

$$\frac{dy}{dx} = 2x + 5$$

Equation of the tangent line

At point $x = 2$

The gradient of the tangent line $m = \frac{dy}{dx} = 2(2) + 5$

$$= 4 + 5 = 9$$

$$m = 9$$

At co-ordinate points $(2,20)$ the equation of the tangent will be

$$y - y_1 = m(x - x_1)$$

$$y - 20 = 9(x - 2)$$

$$y - 20 = 9x - 18$$

$$y = 9x - 18 + 20$$

$$y = 9x + 2$$

Equation of the Normal line

At point $x = 2$

Where $m = 9$

The gradient of the normal $m_1 = \frac{-1}{\frac{dy}{dx}} = \frac{-1}{9}$

At co-ordinate points $(2,20)$ the equation of the Normal will be

$$y - y_1 = \frac{-1(x - x_1)}{m_1}$$

$$y - 20 = \frac{-1(x - 2)}{9}$$

$$9(y - 20) = -x + 2$$

$$9y - 180 = -x + 2$$

$$9y = -x + 2 + 180$$

$$9y = -x + 182$$

$$9y + x - 182 = 0$$

Examples 2

Find the equation of the tangent to the curve $y = x^2 - x - 2$ at point $(1, -2)$

<u>Solution</u>

$$y = x^2 - x - 2$$

How to find the equation of tangent using derivative of the function.

If $\qquad\qquad y = x^2 - x - 2$

The derivative of function y

$$Gradient = m = \frac{dy}{dx} = 2x - 1$$

At point $(1, -2)$, where $x = 1$

$$m = \frac{dy}{dx} = 2x - 1 = 2(1) - 1 = 2 - 1 = 1$$

$$m = 1$$

<u>Equation of the tangent line</u>

At co-ordinate points $(1, -2)$ where $x_1 = 1$, $y_1 = -2$ the equation of the

tangent will be

$$y - y_1 = m(x - x_1)$$

$$y - (-2) = 1(x - 1)$$

$$y + 2 = x - 1$$

$$y = x - 1 - 2 = x - 3$$

$$y = x - 3$$

Examples 3

Find the equation of the normal to the curve $y = 4x^2 - 3x - 2$ at point $(1, -2)$

<u>Solution</u>

$$y = 4x^2 - 3x - 2$$

How to find the equation of normal using derivative of the function.

If
$$y = 4x^2 - 3x - 2$$

The derivative of function y

$$Gradient = m = \frac{dy}{dx} = 8x - 3$$

At point $(1, -2)$, where $x = 1$

$$m = \frac{dy}{dx} = 8x - 3 = 8(1) - 3 = 8 - 3 = 5$$

$$m = 5$$

<u>*Equation of the Normal line*</u>

At co-ordinate points $(1, -2)$ where $x_1 = 1$, $y_1 = -2$ the equation of the

normal will be

$$y - y_1 = \frac{-1(x - x_1)}{m_1}$$

$$y - (-2) = \frac{-1(x - 1)}{1}$$

$$y + 2 = -x + 1$$

$$y + 2 = -x + 1$$
$$y = -x + 1 - 2$$
$$y = -x - 1$$
$$y + x = -1$$

Exercise 9a

1. Find the equation of the tangent and normal to the curve $y = \dfrac{x^3}{3}$ at the point $\left(-1, \dfrac{1}{3}\right)$.

2. Find the equation of the tangent and normal to the curve $y = 2x^2$ at the point $(1, \ 2)$.

3. Find *the equation of the tangent and normal to the curve* $y = 3x^2 - 2x$ *at the point* $(2, 8)$.

4. *Find the equation of the tangent and normal to the curve* $y = \dfrac{x^3}{2}$ *at the point* $\left(-1, \dfrac{-1}{2}\right).$

5. Find the equation of the tangent and normal to the curve $y = -x^2 + x + 1$ at the point $(-2, -5)$.

Increasing And Decreasing Function

Increasing function

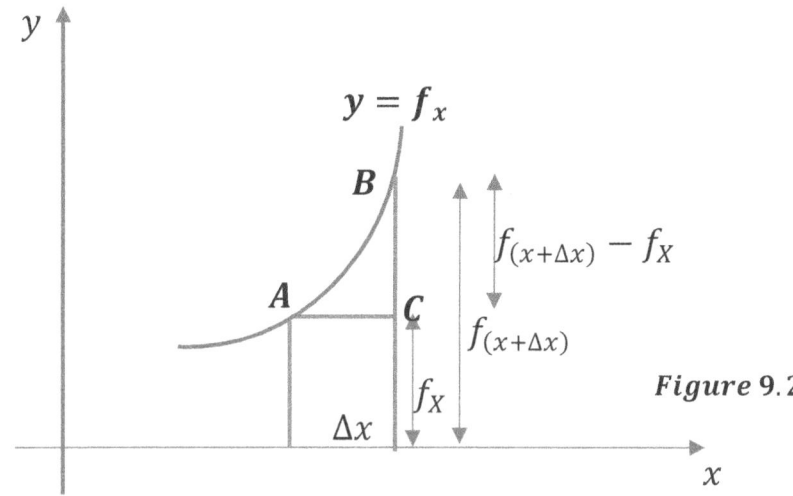

Figure 9.2

From figure 9.2,

$$y = f_x$$

As x $increases, y$ $increases. therefore$ $\frac{dy}{dx} > 0$.

From the definition of $\frac{dy}{dx}$ as a limiting value of $\frac{f_{(x+\Delta x)} - f_x}{\Delta x}$.

If y $increases. therefore$ $\frac{dy}{dx} > 0$, then

$$\frac{f_{(x+\Delta x)} - f_x}{\Delta x} > 0$$

$$f_{(x+\Delta x)} - f_x > 0.$$

Summarily,

If f_x is increasing, then

$$f_{(x+\Delta x)} - f_x > 0.$$

Then the limiting value of

$$\frac{f_{(x+\Delta x)} - f_x}{\Delta x} \ will\ be > 0\ and\ \frac{dy}{dx} > 0$$

Decreasing function

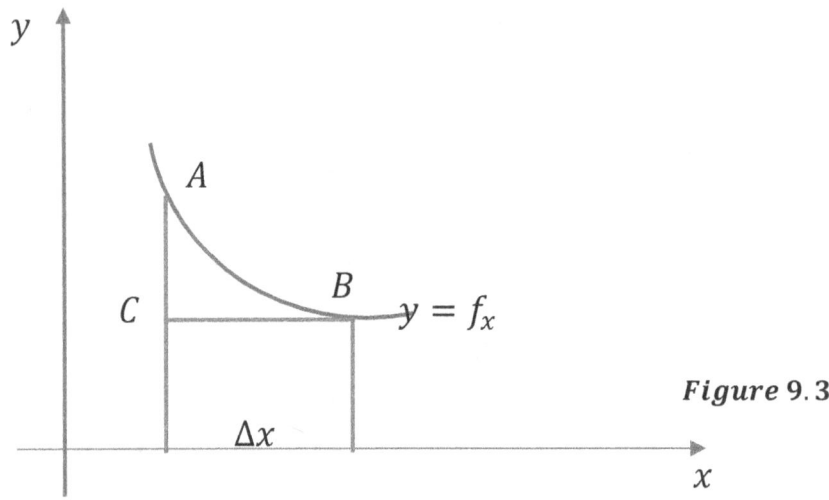

Figure 9.3

From figure 9.3,

$$y = f_x$$

As y *decreases. therefore* $\dfrac{dy}{dx} < 0.$

From the definition of $\dfrac{dy}{dx}$ as a limiting value of $\dfrac{f_{(x+\Delta x)} - f_x}{\Delta x}$.

If y *decreases. therefore* $\dfrac{dy}{dx} < 0$, then

$$\frac{f_{(x+\Delta x)} - f_x}{\Delta x} < 0$$

$$f_{(x+\Delta x)} - f_x < 0.$$

Summarily,

If f_x is a decreasing function, then

$$f_{(x+\Delta x)} - f_x < 0.$$

Then the limiting value of

$$\frac{f_{(x+\Delta x)} - f_x}{\Delta x} \ \textit{will be} < 0 \ \textit{and} \ \frac{dy}{dx} < 0$$

The derivative of a function is positive over the range where it is increasing and negative where it is decreasing.

Examples 1
Find the range of values of x for which $x^2 - 5x$ is increasing

Solution

Let $y = x^2 - 5x$

$$\frac{dy}{dx} = 2x - 5$$

The function y is increasing when $\frac{dy}{dx} > 0$

$$2x - 5 > 0$$

$$2x > 5$$

$$x > \frac{5}{2}$$

Therefore $y = x^2 - 5x$ is increasing as

$$x > \frac{5}{2}$$

$y = x^2 - 5x$ is increasing when $x > \frac{5}{2}$.

Examples 2
Find the range of values of x for which $\frac{x^2}{2} - 7x + 4$ is decreasing

Solution

Let $y = \frac{x^2}{2} - 7x + 4$

$$\frac{dy}{dx} = x - 7$$

The function y is decreasing when $\frac{dy}{dx} < 0$

$$x - 7 < 0$$

$$x < 7$$

$$x < 7$$

Therefore $y = x^2 - 5x$ is decreasing as

$$x < 7$$

Examples 3

Find the range of values of x for which $\frac{x^3}{3} - \frac{x^2}{2} - 2x$ is increasing

Solution

Let $y = \frac{x^3}{3} - \frac{x^2}{2} - 2x$

$$\frac{dy}{dx} = x^2 - x - 2$$

$$= (x + 1)(x - 2)$$

The function y is increasing when $\frac{dy}{dx} > 0$

$$= (x + 1)(x - 2) > 0$$

$$(x + 1) > 0 \quad or \quad (x - 2) > 0$$

$$x < -1 \qquad or \qquad x > 2$$

Therefore $y = \frac{x^3}{3} - \frac{x^2}{2} - 2x$ is increasing as $x < -1$ or $x > 2$

Examples 4

Find the range of values of x for which $2x^2 + 3x + 1$ is decreasing

Solution

Let $y = 2x^2 + 3x + 1$

$$\frac{dy}{dx} = 4x + 3$$

The function y is decreasing when $\frac{dy}{dx} < 0$

$$4x + 3 < 0$$

$$x < -\frac{3}{4}$$

Therefore $y = 2x^2 + 3x + 1$ is increasing as

$$x < -\frac{3}{4}$$

Examples 5

Find the range of values of x for which $x^2 - 4x + 3$ is increasing

Solution

Let $y = x^2 - 4x + 3$

$$\frac{dy}{dx} = 2x - 4$$

The function y is increasing when $\frac{dy}{dx} > 0$

$$2x - 4 > 0$$

$$2x > 4$$

$$x > \frac{4}{2}$$

$$x > 2$$

Therefore $y = x^2 - 4x + 3$ is increasing as

$$x > 2$$

Exercise 9b

1. *Find the range of values of x for which $5 - 3x + x^2$ is increasing*

2. *Find the range of values of x for which $8x - x^2 + \dfrac{x^3}{3}$ is increasing*

3. *Find the range of values of x for which $1 + 4x - 2x^2$ is decreasing*

4. *Find the range of values of* x *for which* $\dfrac{x^3}{3} + \dfrac{x^2}{2} - 6x$ *is decreasing*

5. *Find the range of values of x for which $\dfrac{2x^3}{3} + \dfrac{x^2}{2} - 3x$ is increasing*

Approximation

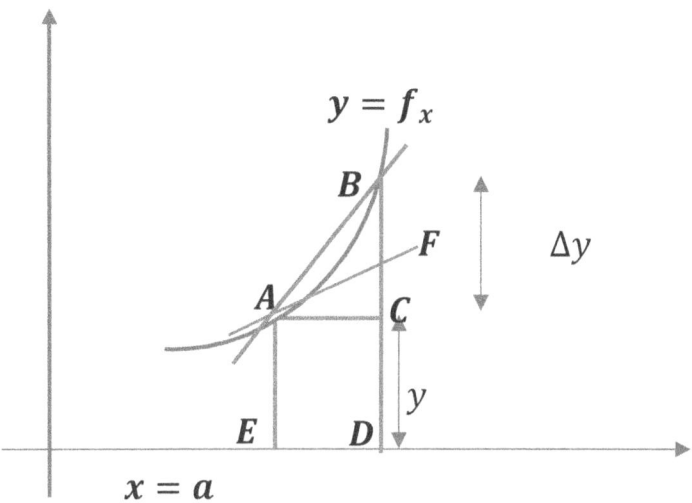

From the curve $y = f_x$

Gradient of secant $AB = \dfrac{\Delta y}{\Delta x}$

At point $x = a$

Gradient of the tangent at $A \dfrac{dy}{dx}$ at $x = a$.

In the figure AB represents a portion of the graph of y. The point A represents the value of y with the given value $x = a$. $AC = \Delta x$, $BC = \Delta y$. However, if Δx *is small we may take* $CF = \Delta y$, where F lies on the tangent at A. The gradient of this tangent is given by the value of $\frac{dy}{dx}$ when $x = a$, which we will write as $\left(\frac{dy}{dx}\right)_{x=a}$.

Now $\frac{CF}{AB} = \left(\frac{dy}{dx}\right)_{x=a}$

Therefore,

$$\frac{\Delta y}{\Delta x} \cong \frac{dy}{dx}$$

$$\Delta \mathbf{y} = \frac{\mathbf{dy}}{\mathbf{dx}} \times \Delta \mathbf{x}$$

Examples 1
If a side of a square increases by 0.5%, find the approximate percentage increase in the area.

Solution

$$Area\ of\ a\ square\ A = l^2$$

The derivative of A with respect to l

$$\frac{dA}{dl} = 2l$$

Percentage increase of square $= \frac{\Delta l}{l} = 0.5\% = \frac{0.5}{100}$

Percentage increase in the Area $= \frac{\Delta A}{A} \times 100$

If

$$\frac{\Delta A}{\Delta l} = \frac{dA}{dl}$$

Therefore

$$\Delta A = \frac{dA}{dl} \times \Delta l$$

Where $\frac{dA}{dl} = 2l$

$$\Delta A = 2l \times \Delta l = 2l\Delta l$$

Percentage increase in the Area $= \frac{\Delta A}{A} \times 100$

$$= \frac{2l\Delta l}{A} \times 100$$

Recall $A = l^2$

$$= \frac{2l\Delta l}{l^2} \times 100$$

$$= \frac{2\Delta l}{l} \times 100$$

If $\frac{\Delta l}{l} = 0.5\% = \frac{0.5}{100}$

$$= 2 \times \frac{0.5}{100} \times 100 = 2 \times 0.5 = 1\%$$

Percentage increase in the Area= 1%

Examples 2

If the radius of a sphere decrease by 0.1%. find the percentage decrease in the
   a. Surface area.
   b. Volume

Solution

Percentage decrease of radius $= \frac{\Delta r}{r} = -0.1\% = \frac{-0.1}{100}$

   a. Surface Area

Surface area of a sphere $S = 4\pi r^2$

The derivative of S with respect to r

$$\frac{dS}{dr} = 8\pi r$$

If $\frac{\Delta S}{\Delta r} \approx \frac{dS}{dr}$

Therefore

$$\Delta S = \frac{dS}{dr} \times \Delta r$$

Percentage decrease in Surface Area $= \frac{-\Delta s}{S} \times 100$

Where $S = 4\pi r^2$, $\Delta S = \frac{dS}{dr} \times \Delta r$, $\frac{dS}{dr} = 8\pi r$

$$= \frac{-\Delta s}{S} \times 100$$

$$= \frac{-\frac{dS}{dr} \times \Delta r}{4\pi r^2} \times 100$$

$$= \frac{-8\pi r \times \Delta r}{4\pi r^2} \times 100$$

$$= \frac{-2 \times \Delta r}{r} \times 100$$

Where $\frac{\Delta r}{r} = \frac{-0.1}{100}$

$$= -2 \times \frac{-0.1}{100} \times 100 = 0.2\%$$

Percentage decrease in Surface Area is 0.2%

    b.   *Volume*

Volume of a sphere $V = \frac{4}{3}\pi r^3$

The derivative of V with respect to r

$$\frac{dV}{dr} = 4\pi r^2$$

If $\frac{\Delta V}{\Delta r} \approx \frac{dV}{dr}$

Therefore

$$\Delta V = \frac{dV}{dr} \times \Delta r$$

Percentage decrease in Volume $= \dfrac{-\Delta V}{V} \times 100$

Where $V = \dfrac{4}{3}\pi r^3$, $\Delta V = \dfrac{dV}{dr} \times \Delta r.$ $\dfrac{dV}{dr} = 4\pi r^2$

$$= \dfrac{-\Delta V}{V} \times 100$$

$$= \dfrac{-\dfrac{dS}{dr} \times \Delta r}{\dfrac{4}{3}\pi r^3} \times 100$$

$$= \dfrac{-4\pi r^2 \times \Delta r}{\dfrac{4}{3}\pi r^3} \times 100$$

$$= \dfrac{-3 \times \Delta r}{r} \times 100$$

Where $\dfrac{\Delta r}{r} = \dfrac{-0.1}{100}$

$$= -3 \times \dfrac{-0.1}{100} \times 100 = 0.3\%$$

Percentage decrease in Volume is 0.3%

Rate Of Change

If $y = f_{(x)}$, $\dfrac{dy}{dx}$ can sometimes be interpreted as the rate at which y changes with respect to x. If y increases as x increases $\dfrac{dy}{dx} > 0$. While if y decreases as x increases $\dfrac{dy}{dx} < 0$.

Examples 1
The radius of a circle is increasing at the rate of $0.1\ cm/s$. find the rate at which the area is increasing when the radius oif the circle is $5cm$

Solution

The rate of increase of the radius of a circle $\frac{dr}{dt} = 0.1 \, cm/s$

Area of a cicle A= πr^2

Differentiate A wrt r

$$\frac{dA}{dr} = 2\pi r$$

Using chain rule

$$\frac{dA}{dt} = \frac{dA}{dr} \times \frac{dr}{dt}$$

$$\frac{dA}{dt} = 2\pi r \times 0.1$$

Where $r = 5cm \; \pi = 3.142$

$$\frac{dA}{dt} = 2 \times 3.142 \times 5 \times 0.1$$

$$\frac{dA}{dt} = 10 \times 3.142 \times 0.1 = 3.142$$

$$\frac{dA}{dt} = 3.142 \, cm^2/s$$

Examples 2

Water is leaking from a hemispherical bowl of radius 30cm at the rate of $0.5 \, cm^2/s$.find the rate at which the surface area of the water is decreasing when the water level is half way from the top.

Solution

Area of the surface of the hemisphere bowl fo radius r $(A) = \pi r^2$

$$\frac{dA}{dr} = 2\pi r$$

$$\frac{dA}{dt} = 0.5 \, cm^2/s$$

when the water level is half way

$$V = \frac{\frac{4}{3}\pi r^3}{2}$$

Using chain rule

$$\frac{dA}{dt} = \frac{dA}{dr} \times \frac{dr}{dt}$$

$$\frac{dA}{dt} = 2\pi r \times 0.1$$

Where $r = 5cm$ $\pi = 3.142$

$$\frac{dA}{dt} = 2 \times 3.142 \times 5 \times 0.1$$

$$\frac{dA}{dt} = 10 \times 3.142 \times 0.1 = 3.142$$

$$\frac{dA}{dt} = 3.142 \, cm^2/s$$

Examples 2

Water is leaking from a hemispherical bowl of radius 30cm at the rate of $0.5 \, cm^2/s$.find the rate at which the surface area of the water is decreasing when the water level is half way from the top.

Solution

Area of the surface of the hemisphere bowl fo radius r $(A) = \pi r^2$

$$\frac{dA}{dr} = 2\pi r$$

$$\frac{dA}{dt} = 0.5 \, cm^2/s$$

when the water level is half way

$$V = \frac{\frac{4}{3}\pi r^3}{2}$$

Using chain rule

$$\frac{dA}{dt} = \frac{dA}{dr} \times \frac{dr}{dt}$$

$$\frac{dA}{dt} = 2\pi r \times 0.1$$

Where $r = 5cm$ $\pi = 3.142$

$$\frac{dA}{dt} = 2 \times 3.142 \times 5 \times 0.1$$

$$\frac{dA}{dt} = 10 \times 3.142 \times 0.1 = 3.142$$

$$\frac{dA}{dt} = 3.142 \, cm^2/s$$

Examples 3

The length l meters of a metal rod at temperature $\theta^0 C$ is given by $l = 1 + 0.05\theta + 0.04\theta^0$. Determine the rate of change of length, when the temperature is a. $200^0 C$ b. $300^0 C$

Solution

The length of the rod $l = 1 + 0.05\theta + 0.04\theta^0$

$$Rate \; of \; change \; of \; length \; to \; temperature = \frac{dl}{d\theta} = 0.05 + 0.08\theta$$

a. when $\theta = 200^0 C$

$$\frac{dl}{d\theta} = 0.05 + 0.08(200)$$

$$\frac{dl}{d\theta} = 0.05 + 16 = 16.05 \, m/^0 C$$

b. when $\theta = 300^0 C$

$$\frac{dl}{d\theta} = 0.05 + 0.08(300)$$

$$\frac{dl}{d\theta} = 0.05 + 24 = 24.05 \ m/^0C$$

Examples 4

The area of a circle is increasing at the rate of $4cm^2s^{-1}$. Find the rate of change of the circumference when the radius is 6cm.

Solution

$$Radius \ of \ the \ circle \ (r) = \ 6cm$$

$$Area \ of \ a \ circle \ (A) = \ \pi r^2$$

$$Rate \ of \ change \ of \ Area \ to \ radius = \frac{dA}{dr} = 2\pi r$$

$$Rate \ at \ which \ Area \ Increases = \frac{dA}{dt} = 4\text{cm}^2\text{s}^{-1}$$

$$Circumference \ of \ a \ circle \ (C) = \ 2\pi r$$

$$Rate \ of \ change \ of \ Circumference \ to \ radius = \frac{dC}{dr} = 2\pi$$

$$Rate \ of \ change \ of \ Circumference \ \frac{dC}{dt} = \frac{dC}{dr} \times \frac{dA}{dt} \times \frac{1}{\frac{dA}{dt}}$$

$$\frac{dC}{dt} = 2\pi \times 4 \times \frac{1}{2\pi r} = \frac{4}{r} = \frac{4}{6} = \frac{2}{3} cms^{-1}$$

Examples 5

Suppose the radius (r) of a circle is $3cm$ at a certain instant, and it is increasing at the rate of $0.5cms^{-1}$. At what rate will the area (A) be increasing at that instant.

Solution

$$Radius \ of \ the \ circle \ (r) = \ 3cm$$

$$Rate\ of\ increase\ in\ radius = \frac{dr}{dt} = 0.5 cms^{-1}$$

$$Area\ of\ a\ circle\ (A) = \pi r^2$$

$$Rate\ of\ change\ of\ Area\ to\ radius = \frac{dA}{dr} = 2\pi r$$

$$Rate\ at\ which\ Area\ Increases = \frac{dA}{dt} = \frac{dA}{dr} \times \frac{dr}{dt} = 2\pi r \times 0.5$$

$$= 2\pi \times 3 \times 0.5 = 3\pi$$

$$where\ \pi = 3.142$$

$$\frac{dA}{dt} = 3 \times 3.142 = 9.426$$

Exercise 9c

1. At what rate is the area of a circle decreasing when its radius is 4cm, and decreasing at the rate of $0.2cms^{-1}$

2. *when the circumference of a circle is 250mm, the radius is increasing at rate of $5mms^{-1}$, find the rate at which area of the circle is increasing.*

3. The angle of a sector of a circle is 75^0. If the radius of this circle is increasing at the rate of $0.3cms^{-1}$. Find the rate of increase of the area when the radius is $6cm$, Leaving your answer in terms of π

4. The luminous intensity l candelas of a lamp at varying voltage V is given by: $l = 4 \times 10^{-4}V^2$. determine the voltage at which the light is increasing at a rate of 0.6 candela per volt.

5. *The radius "r" of a sphere is 4cm and is increasing at the rate of* $0.5cms^{-1}$. *At what rate is the volume "V" increase if* $V = \frac{4}{3}\pi r^3$

Rectilinear Motion

This is a motion of a particle along a straight line. It is specified by the equation $x = f_{(t)}$. Where x is the distance of the particle, from initial point O and t is the time. We have a displacement if a direction is associated with distance.

If V is the velocity, the velocity, the velocity of a particle P is the rate of change of displacement x.

Note:

1. If the body is at rest $\frac{dx}{dt} = 0$

2. If the particle moves towards O, the initial point $\frac{dx}{dt} < 0$

3. If the particle moves away from 0, the initial point, $\frac{dy}{dx} > 0$.

At distance x at time t

Velocity= $\frac{dx}{dt}$

Acceleration $(a) = \frac{d}{dt}(V) = \frac{dv}{dt}$

$$= \frac{d}{dt}\left(\frac{dx}{dt}\right)$$

$$= \frac{d^2x}{dt^2}$$

- At increasing velocity $\frac{dv}{dt} > 0$ and the body is said to be accelerating.

- At decreasing velocity $\frac{dv}{dt} < 0$ and the body is said to be decelerating

.

- At constant speed $\frac{dv}{dt} = 0$

Examples 1

The motion of a particular body along a straight line is specified by the equation.$4t^4 - 3t^3$

Find the velocity and acceleration after 3 seconds.

Solution

A_____ x _____ B

Velocity of the Body

$$Velocity = \frac{displacement\ (x)}{time\ (t)}$$

$$V = \frac{dx}{dt}$$

If $x = 4t^4 - 3t^3$

$$\frac{dx}{dt} = 16t^3 - 9t^2$$

At time $t = 3s$

$V = 16(3^3) - 9(3^2) = (16 \times 27) \times (9 \times 9)$

$$= 432 - 81 = 351\ m/s$$

the velocity of the body along the straight line is $351\ m/s^2$

Acceleration of the Body

$$Acceleration\ (a) = \frac{Velocity\ (x)}{time\ (t)} = \frac{dv}{dt} = \frac{d^2x}{dt^2}$$

$$a = \frac{dv}{dt} = \frac{d^2x}{dt^2}$$

If $\frac{dx}{dt} = 16t^3 - 9t^2$

$$a = \frac{dv}{dt} = \frac{d^2x}{dt^2} = 48t^2 - 18t$$

At time $t = 3s$

$$a = \frac{dv}{dt} = 48 \times 3^2 - 18 \times 3 = 432 - 54 = 378 \, m/s^2$$

Examples 2

The distance x metres moved by a car in a time t seconds is given by $x = 3t^3 - 2t^2 + 4t - 1$. Determine the velocity and acceleration when a. $t = 0$ and b. $t = 2s$

<u>Solution</u>

$$x = 3t^3 - 2t^2 + 4t - 1$$

Velocity of the Body

$$Velocity = \frac{displacement\ (x)}{time\ (t)}$$

$$V = \frac{dx}{dt}$$

If $x = 3t^3 - 2t^2 + 4t - 1$

$$\frac{dx}{dt} = 9t^2 - 4t + 4$$

At time $t = 0$

$V = 9(0)^2 - 4(0) + 4$

$$= 0 - 0 + 4 = 4 \, m/s$$

$$the\ velocity\ of\ the\ car\ 4\ m/s$$

At time $t = 2$

$V = 9(2)^2 - 4(2) + 4 = 9(4) - 4(2) + 4$

$$= 36 - 8 + 4 = 32 \, m/s$$

the velocity of the car $32\,m/s$

Acceleration of the Body

$$Acceleration\,(a) = \frac{Velocity\,(x)}{time\,(t)} = \frac{dv}{dt} = \frac{d^2x}{dt^2}$$

$$a = \frac{dv}{dt} = \frac{d^2x}{dt^2}$$

If $\frac{dx}{dt} = 9t^2 - 4t + 4$

$$a = \frac{dv}{dt} = \frac{d^2x}{dt^2} = 18t^2 - 4$$

At time $t = 0$

$$a = \frac{dv}{dt} = 18(0)^2 - 4 = -4 = -4\,m/s^2$$

At time $t = 2$

$$a = \frac{dv}{dt} = 18(2)^2 - 4 = 18 \times 4 - 4$$

$$= 72 - 4 = 68\,m/s^2$$

Examples 3

The angular displacement θ radians of a flywheel varies with time t second and follows the equation $\theta = 9t^2 + 2t^3$. Find the angular velocity and acceleration of the flywheel when $t = 1s$.

Solution

$$\theta = 9t^2 + 2t^3$$

Angular Velocity of the Body ω

$$Angular\ Velocity\ \omega = \frac{Angular\ displacement\ (\theta)}{time\ (t)}$$

$$\omega = \frac{d\theta}{dt}$$

If

$$\theta = 9t^2 + 2t^3$$

$$\frac{d\theta}{dt} = 18t - 6t^2$$

At time $t = 1$

$$\omega = 18t - 6t^2 = 18(1) - 6(1)^2$$

$$= 18 - 6 = 12\ m/s$$

$$the\ angular\ velocity\ of\ the\ car\ 12\ rad/s$$

Angular Acceleration of the Body

$$Angular\ Acceleration\ (a) = \frac{Angular\ Velocity\ (\omega)}{time\ (t)} = \frac{d\omega}{dt}$$

$$= \frac{d^2x}{dt^2}$$

$$a = \frac{d\omega}{dt} = \frac{d^2\theta}{dt^2}$$

If $\frac{d\theta}{dt} = 18t - 6t^2$

$$a = \frac{d\omega}{dt} = \frac{d^2\theta}{dt^2} = 18 - 12t$$

At time $t = 1$

$$a = \frac{dv}{dt} = 18 - 12(1) = 18 - 12 = 6\,rad/s^2$$

At time t = 2

$$a = \frac{dv}{dt} = 18(2)^2 - 4 = 18 \times 4 - 4$$

$$= 72 - 4 = 68\,m/s^2$$

Examples 4

A body is projected vertically upward and the height hm reached after a time ts, is given by $h = 196t - 4.9t^2$, Find;

a. The time taken to reach the greatest height.

b. The greatest height reached.

Solution

$$h = 196t - 4.9t^2$$

a. Timetaken to reach the greatest height.

$$B\,\bigg|$$
$$h = 196t - 4.9t^2$$
$$A\,\bigg|$$

At the height B, final velocity V=0

$$Velocity, V = \frac{displacement\,(height)}{time\,taken}$$

Considering the given function of height h,

$$V = \frac{dh}{dt} = 196 - 9.8t$$

At $v = 0$

$$196 - 9.8t = 0$$

$$196 = 9.8t$$

divide both sides by 9.8

$$\frac{196}{9.8} = \frac{9.8t}{9.8}$$

$$20s = t$$

Therefore the timetaken to reach the greatest height is 20s

b. Greatest height reached.

At $t = 20s$

$$h = 196t - 4.9t^2$$

$$h = 196 \times 20 - 4.9 \times 20^2 = 3920 - 1960 = 1960m$$

$$\boldsymbol{h = 1960m}$$

Exercise 9d

1. A missile fired from ground level rises $x \; metres$ vertically upward t seconds and $x = 100t - \frac{15}{2}t^2$. Find the velocity of the missile at time $t = 0$

2. The motion of a body is describe to move a distance $s = \frac{2}{3}t^3 - \frac{17}{2}t^2 + 21t$, which is measured in metres. Find the acceleration of the body when it is momentarily at rest.

3. The displacement $x cm$ of a particle is given by; $x = 2.2 \cos 5\pi t + 3.6 \sin 5\pi t$. Evaluate the veocity in m/s when time $t = 0.003s$

4. A particle was dropped from a cliff and the distance fallen in time t seconds is given by: $x = \frac{1}{2}gt^2 m/s^2$. Find the velocity and acceleration of the particle after fallen for $3s$.

STATIONARY POINT

Stationary point is the point on the curve at which $\frac{dy}{dx} = 0$. The value of the function represented by the curve at that point is called its stationary value. Stationary point is applied at $\frac{dy}{dx} = 0$, to determine the value at maximum points, minimum point and the point of inflexion which will be explained later. Maximum and minimum points are also called turning points.

Examples 1

Find the stationary points of the curve which equation is.

$$4x^3 + 15x^2 - 18x + 7$$

Solution

To determine the stationary points of the function, the derivative of the function will be equated to 0

Let $y = 4x^3 + 15x^2 - 18x + 7$

Derivative of y

$$\frac{dy}{dx} = 12x^2 + 30x - 18$$

Stationary points, $\frac{dy}{dx} = 0$

$$12x^2 + 30x - 18 = 0$$
$$6(2x^2 + 5x - 3) = 0$$

Solve the quadratic function;

$$6(2x^2 + 5x - 3) = 0$$
$$6(2x^2 + 6x - x - 3) = 0$$
$$6\big(2x(x + 3) - 1(x + 3)\big) = 0$$

$$6\big((2x - 1)(x + 3)\big) = 0$$

$$(2x - 1)(x + 3) = 0$$

$$2x - 1 = 0 \ or \ x + 3 = 0$$

$$x = \frac{1}{2} \ or \ x = -3$$

Therefore the stationary points is at $x = \frac{1}{2}$ or $x = -3$

Examples 2

Find the stationary points of the curve which equation is;

$$\frac{x^3}{3} + x^2 - 3x + 7$$

Solution

To determine the stationary points of the function, the derivative of the function will be equated to 0

Let $y = \frac{x^3}{3} + x^2 - 3x + 7$

Derivative of y

$$\frac{dy}{dx} = x^2 + 2x - 3$$

Stationary points, $\frac{dy}{dx} = 0$

$$x^2 + 2x - 3 = 0$$

$$(x^2 + 2x - 3) = 0$$

Solve the quadratic function;

$$(x^2 + 2x - 3) = 0$$

$$(x^2 + 3x - x - 3) = 0$$

$$x(x + 3) - 1(x + 3) = 0$$
$$(x - 1)(x + 3) = 0$$

$$x - 1 = 0 \ or \ x + 3 = 0$$

$$x = 1 \ or \ x = -3$$

Therefore the stationary points is at $x = 1$ or $x = -3$

Maximum And Minimum Points

Maximum point

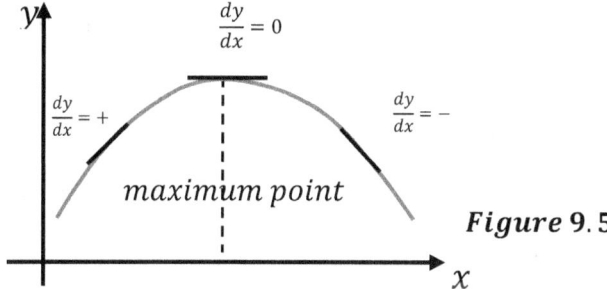

Figure 9.5

Figure 9.5 shows a curve passing through a stationary point and reaching a maximum value at that point. As x increases the gradient of the curve decreases from a positive value through 0 to a negative value.

Consider the change in value of $\frac{dy}{dx}$ around a maximum point. $\frac{dy}{dx}$ is decresing as it goes from +ve values. Hence $\frac{dy}{dx}$ is a decreasing function at this point and its derivative, $\frac{d^2y}{dx^2} < 0$. Hence the necessary and sufficient conditions for a maximum point at $x = a$ on the curve $y = f_x$ are;

i. $\frac{dy}{dx} = 0$

ii. $$\frac{d^2y}{dx^2}_{x=a}$$

Minimum Point

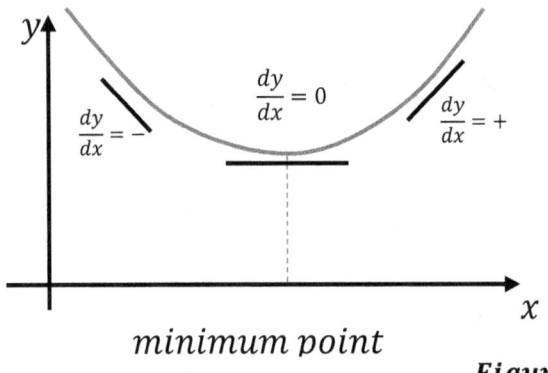

minimum point

Figure 9.6

Figure 9.6 shows a minimum value at a stationary point. As x increases along the curve the gradient increases from a negative value through 0 then a positive (+ve) value.

Maximum and minimum points are also called **turning points**, as the tangent turns around at such points.

At a minimum point, $\frac{dy}{dx}$ increases from -ve values to +ve values. From the figure above $\frac{dy}{dx}$ is an increasing function at this point and so $\frac{d^2y}{dx^2} > 0$.

Summarily, for maximum and minimum point, $\frac{dy}{dx}$ must equal 0. For maximum point $\frac{d^2y}{dx^2} < 0$, and for a minimum point $\frac{d^2y}{dx^2} > 0$.

Point Of Inflexion

In this case the curve has neither maximum nor minimum value but $\frac{dy}{dx} = 0$. This is called a point of inflexion, that is, this is thepoint where the targent "bends" or flexes aand then continues +ve or -ve as it was.

Along the curve of the figure below Op and On are points of inflexion

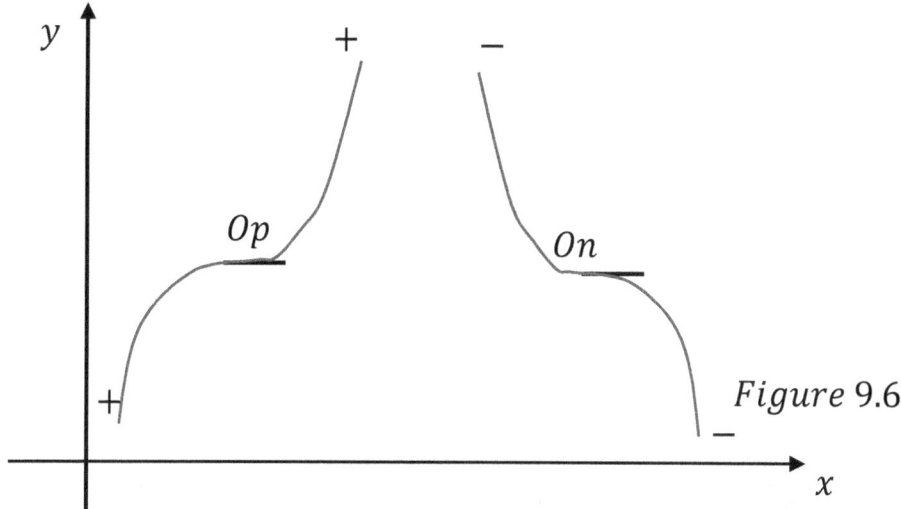

Figure 9.6

TEST FOR STATIONARY POINTS USING SECOND DERIVATIVE

A polynomial function of the second degree (a quadratic function) will give a gradient function of the first degree. Hence, it can only have a one stationary point, which must either be a maximum or minimum. The graph of the function is parabola, which has one turning point.

For a curve, the necessary stationary point of curve $y = f_x$ is determine when $\frac{dy}{dx} = 0$. This is not enough to determine the nature of the stationary points. In this section, we shall look into the second derivative testfor stationary points.

Steps for finding and differentiating between stationary points;

1. Given a function $y = f_x$; determine $\frac{dy}{dx}$.

2. Equate $\frac{dy}{dx} = 0$, then find the value of x.

3. Substitute the value of x into the original function $y = f_x$, to find the corresponding $y - coordinate$ value. Hence, {x,y} are the ordinates of the stationary points.

To determine the nature of the stationary point:

4. Find $\frac{d^2y}{dx^2}$, then substitute the value of x gotten in option (2) into it. If after substitution, the answer is;

   a. *Positive*: the point is a minimum point.

   b. *Negative*: the point is a maximum point.

   c. *Zero*: the point is a point of inflexion

Or

5. Determine the sign of the gradient of the curve just before and after the stationary points. If the changes of sign for the gradient of the curve is;

   a. *Positive to negative:* the point is a maximum point.

   b. *Negative to positive*: the point is a minimum point.

   c. *Positive to positive or negative to negative*: the point is a point of inflexion.

Examples 1

Locate the tuning points on the curve $y = 4x^2 - 8x$ and determine its nature by examining the sign of the gradient on either side

Solution

$$y = 4x^2 - 8x$$

Following the steps stated above;

1. Determine $\frac{dy}{dx}$.

If $y = 4x^2 - 8x$, $\frac{dy}{dx} = 8x - 8$

2. Equate $\frac{dy}{dx} = 0$, then solve for x

$$8x - 8 = 0$$
$$8x = 0 + 8$$
$$8x = 8$$
$$x = 1$$

3. Substitute $x = 1$ into the original function,

$$y = 4x^2 - 8x$$
$$y = 4(1)^2 - 8(1) = 4 - 8 = -4$$

Therefore; $\{x, y\} = \{1, -4\}$

Hence, the co-ordinates of the turning point is $\{1, -4\}$

To determine the nature of the stationary points, the value of x is slightly reduced by 0.1 and increased by 0.1.

When $x = 0.9$

$$\frac{dy}{dx} = 8x - 8 = 8(0.9) - 8 = 7.2 - 8 = -0.8 \ (negative)$$

When $x = 1.1$

$$\frac{dy}{dx} = 8x - 8 = 8(1.1) - 8 = 8.8 - 8 = 0.8 \ (positive)$$

From the result, the gradient of the curve is negative just before the turning poing and positive after the turning point, i.e − U +.

Hence, the co-ordinates of the turning point is {1, −4} is a **_minimum point._**

Exercise 9e

1. Locate the turning point on the following curve and determine, whether it is a maximum or minimum point: $y = 4\theta + e^{-\theta}$

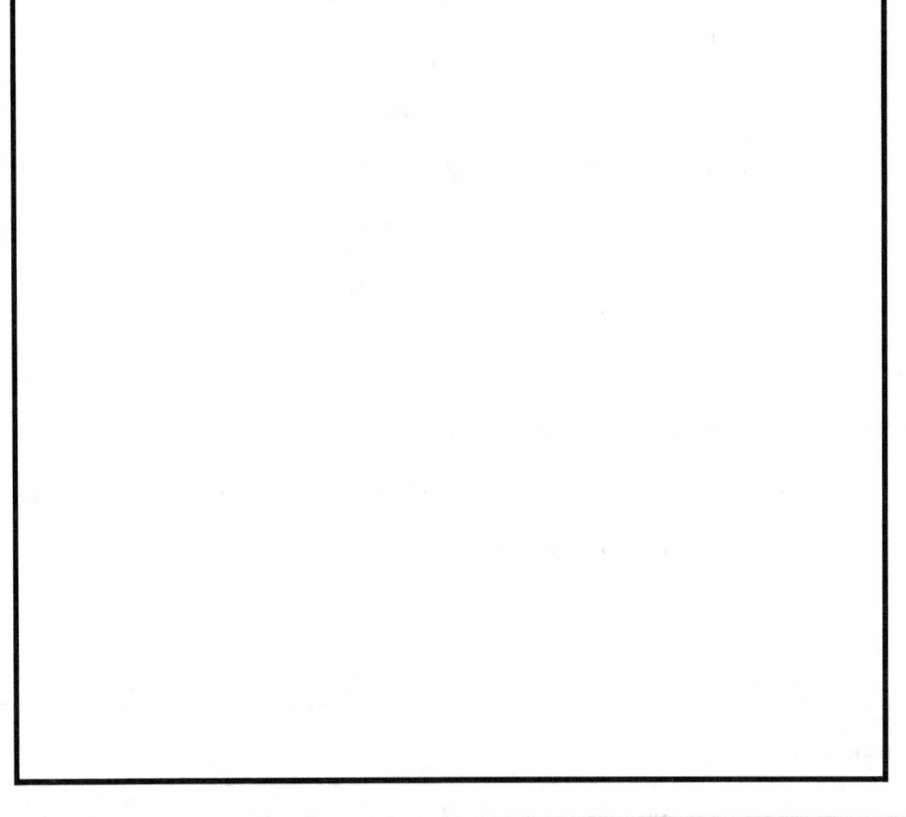

2. In $y = 3x^2 - 4x + 2$, Find the turning points and distinguish between them

3. In $y = 5x - 2 \ln x$, Find the turning points and distinguish between them.

4. Determine the co-ordinates of the maximum and minimum values of the graph $y = \dfrac{x^3}{3} - \dfrac{x^2}{2} - 6x + \dfrac{5}{3}$

Word problem on maximum and minimum value.

Examples 1
Find the greatest product of two numbers, whose sum is 12.
$$4x^3 + 15x^2 - 18x + 7$$

Solution

Let the number be x and $12 - x$

Let y be the product of the two numbers, then

$$y = x(12 - x)$$

$$= 12x - x^2$$

$$\frac{dy}{dx} = 12 - 2x$$

At the maximum value, $\frac{dy}{dx} = 0$

$$12 - 2x = 0$$

$$2(6 - x) = 0$$

$$(6 - x) = 0$$

$$x = 6$$

Hence $x = 6$, gives a maximum value of y.

Therefore;

$$y = x(12 - x) = 6(12 - 6) = 6 \times 6 = 3$$

Examples 2
100m of wire is available for fencing a rectangular piece of land. Find the dimensions of the land which maximise the area. Hence, determine the maximum area of the fence.

Solution

Let the length of one side of the fence be x then the length of the other side is $\frac{100-2x}{2} = 50 - x$.

Let y be the area of the land, then

$$y = x(50 - x)$$
$$= 50 - x^2$$
$$\frac{dy}{dx} = 50 - 2x$$

$50 - x$

x x

$50 - x$

At the maximum value, $\frac{dy}{dx} = 0$

$$50 - 2x = 0$$
$$2x = 50$$
$$x = 25$$

Hence, the dimension of the land that gives maximum area are $25m \times 25m$

$$Maximum\ area = (25 \times 25)m^2 = 625m^2$$

Exercise 9f

1. The speed, v, of a car in m/s is related to time ts by equation $v = 3 + 12t - 3t^2$. Determine the maximum speed of the car in $\frac{km}{h}$.

2. Determine the area of the largest piece of rectangular ground that can be enclosed by $100m$ of fencing, if part of an exising straight wall is used as one side.

3. Find the height and radius of a closed cylinder of volume $125cm^2$ which has the least surface area.

4. Determine the maximum area of a rectangular piece of land that can be enclosed by $1200m$ of fencing

ANSWERS

EXERCISE 1

1. -7
2. -2
3. 5
4. -10
5. -7
6. $\frac{1}{2}(a^2 - 2a - 5)$

EXERCISE 2

1. $-3 + a$
2. 4
3. $\frac{20}{11}$
4. $\frac{1}{3}$
5. 4
6. $\frac{2}{5}$
7. 518
8. 17
9. 6
10. 30

EXERCISE 3

1. $f(x)$ is continous
2. $h(x)$ is continous

EXERCISE 4

1. $\frac{-3}{x^2}$
2. $4x - 1$
3. $1 + \frac{1}{x^2}$
4. $10x$
5. $6x$
6. $12x^2 - 6x - 5$
7. $7x^6$
8. $35x^4$
9. $\frac{-18}{x^3}$
10. $\frac{1}{2\sqrt{x}} - \frac{1}{2(\sqrt{x})^3}$

EXERCISE 6a

1. $8(-3 + a)^3$
2. $-18x(-3 + a)^3$
3. $2(4x - 3)^{-\frac{1}{2}}$
4. $-3(6x - 5)^{-\frac{3}{2}}$
5. $-3\left(1 + \frac{1}{x^2}\right)\left(x - \frac{1}{x}\right)^2$

EXERCISE 6b

1. $6(x + 4)(x + 1)$
2. $(x^2 - 1)^2(7x^2 - 1)$
3. $x(5x^2 - 3x + 1)$
4. $3x^2(x^2 + 4)(7x^2 + 12)$
5. $(x + 3)(5x + 3)/2\sqrt{x}$

EXERCISE 6c

1. $2(x+1)^{-3}$

2. $\dfrac{(x^2-2x-1)}{x-1}$

3. $\dfrac{(x-2)}{2(x-1)^{\frac{3}{2}}}$

4. $\dfrac{(2x^2-3)^2(10x^2-3)}{x^2}$

5. $\dfrac{-(x+2)}{2(x-2)^2}$

EXERCISE 6d

1. $\dfrac{2x-y}{x}$

2. $\dfrac{-2x}{3y^2}$

3. $\dfrac{2x+3y}{7-3x}$

4. $\dfrac{2x(1-y)}{x^2+1}$

5. $\dfrac{3y-x^2}{y^2-3x}$

EXERCISE 7a

1. $\dfrac{\cos\sqrt{x}}{2\sqrt{x}}$

2. $\cos x - x\sin x$

3. $-\csc^2 2 + x$

4. $8x\sec^2 4x^2$

5. $\dfrac{\sec x}{x}\left(\tan x - \dfrac{1}{x}\right)$

EXERCISE 7b

1. $\dfrac{4}{1+x}$

2. $\dfrac{\frac{1}{x}(1+\sin x)-\cos x\log_e x}{(1+\sin x)^2}$

3. $x(1+2\log_e x)$

4. $32(1+2x)^{-3}$

5. $\dfrac{1}{x}\log_a e$

EXERCISE 7c

1. $-2(e)^{-2x-3}$

2. $\dfrac{e^{\sqrt{x}}}{2\sqrt{x}}$

3. $2x(x+1)e^{2x}$

4. $(\cos x + $
   $ -\sin x)e^{\sin x-\cos x}$

5. $2(x+3)e^{(x+3)^2}$

EXERCISE 8

1. $\dfrac{dy}{dx}=3x^2+6x, \dfrac{d^2y}{dx^2}=$
   $6x+6, \dfrac{d^3y}{dx^3}=6$

2. $6x\sec^2 x^3 - $
   $18x^4\sec^2 x^3\tan x^3$

3. $\dfrac{dy}{dx}=2xe^{x^2}, \dfrac{d^2y}{dx^2}=$
   $4x^2e^{x^2}+2e^{x^2},$

4. $-4\sin 2x$

5. $\dfrac{2}{(x+1)^3}$

EXERCISE 9a

1. a. $3y - 3x = 4$

    b. $3y + 3x = -2$

2. a. $y = 4x - 2$

    b. $4y + x = 9$

3. a. $y = 10x - 12$

    b. $10y + x = 82$

4. a. $y = \frac{3}{2}x + 1$

    b. $6y + 4x + 7 = 0$

5. a. $y = 5x + 5$

    b. $5y + x + 27 = 0$

EXERCISE 9b

1. $x < \frac{3}{4}$

2. $-2 < x < 4$

3. $x > 1$

4. $-3 < x < 2$

5. $x > 1 \ or \ x < \frac{-3}{2}$

EXERCISE 9c

1. $1.6\pi cm/s$

2. $1250mm^2/s^2$

3. $0.75\pi cm^2/s$

4. $750 volts$

5. $32\pi cm^3/s$

EXERCISE 9d

1. $100 \, m/s$

2. $11 \, m/s^2$

3. $0.347 m/s$

4. $Vel. = 29.4 m/s$

    $Acc. = 9.8 \, m/s^2$

EXERCISE 9e

1. $(-1.3863, -1.5452)$

    Is a minimum point.

2. $minimum \ at \ \left(\frac{2}{3}, \frac{2}{3}\right)$

3. $minimum \ at \ (0.4, 3.8326)$

4. $min. \left(3, 11\frac{5}{6}\right)$

    $max. (-2, 9)$

EXERCISE 9f

1. $54 km/h$

2. $1250 m^2$

3. $height = 5.42 cm$

    $radius = 2.71 cm$

4. $9000 m^2$

About the Author

Samuel Adegboye is a dedicated scientist and lecturer who has devoted a significant portion of his life to academia and practical applications. He has actively assisted individuals of all ages in resolving scientific and mathematical problems. With a background in Electrical/Electronic Engineering, Samuel successfully tackled various science subjects, including Mathematics, Physics, Chemistry, and more, during his academic journey. As the visionary behind Kunlektra Academy, his primary objective is to empower young individuals academically, regardless of their backgrounds. Samuel firmly holds the belief that no subject or topic is inherently challenging; rather, he emphasizes the importance of understanding the fundamentals, as it helps prevent numerous errors along the way.

Acknowledgments

First of all, my appreciation goes to God, Almighty, for the opportunity to collate this manuscript. This book was published, with the support of God Almighty and some persons he used to be part of the success of this book. I am grateful for some friends, colleagues, and co-members for encouragement for the support me to start the work, and finally published it.

THANKS FOR PATRONAGE

Improve your Math Skills with other Books

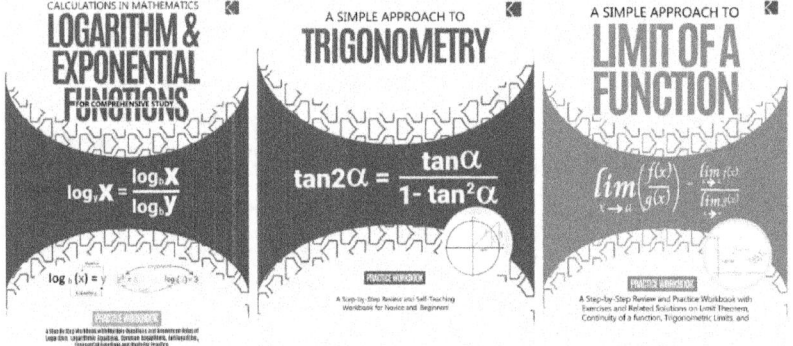

www.ingramcontent.com/pod-product-compliance
Lightning Source LLC
Chambersburg PA
CBHW060917120626
46553CB00001B/361